Necessity and challenge of

Co-operating with Invisible Beings

Daniel Perret

My heartfelt thanks to my teachers, to the spirit beings I have the pleasure to communicate with and to Linda Bloom, Sarah Schwartz, Sarajane Williams and Marie Perret for valuable feedback and for English corrections.

© 2022 Daniel Perret
Publisher: BoD - Books on Demand info@bod.fr
Printing: BoD – Books on Demand,
In de Tarpen 42, Norderstedt (Germany)
Print on demand

Publishing date October 2022
ISBN 978-2-3224-3235-6

BoD has also published this book in German.

Some introductory thoughts
The science of the invisible worlds developed here and there in bits over time. When it comes to the study of nature spirits and their intelligence, less work has been done. Even Rudolf Steiner did not put his main efforts into this. The knowledge of these worlds throughout the millennia existed mainly in what I would call natural folk wisdom and experience, and was rarely the domain of scholars' and what later developed as 'official university science'. All cultures around the world had an intimate, though often mythical intuitive contact to nature and its intelligence. It would not have survived otherwise.

One of the purposes of this book is to look into what the obstacles were that meant that this widespread folk knowledge was rarely considered as and by science.

Times are changing. Is it the scientific tinge of the new Aquarian age that is bringing the tools and focus onto seriously researching the consciousness of the invisible? We can no longer hide it passively behind the concept of a 'God', 'Nature', and 'Fairy Tales' or 'fear of the unknown'. While official science ventures into invisible nuclear energies or quantum physics, the science of subtle energy is developing fast as does the research into consciousness. Wolfgang Weirauch's 1000 published interviews with nature spirits in the German Flensburger Hefte definitely participate in laying the foundation of a new science of the living. Our 'western' scientific approach brings more detailed understanding and discernment. Discernment is becoming of utmost importance. Whom do we have to do with? Who is conveying serious content of communications with nature's intelligent beings? Our present time needs co-operation, and as we realise, we are all connected, yet we need to redouble our efforts to tell friend from foe, truth from illusion.
This is the main purpose of this book and of my work in general.

Table of contents

Continuous texts on the right-hand pages

Daniel Perret – Beings of the Invisible Dimensions
Left side: Conversations with beings

Foreword

After forty years of study and contact with non-physical beings, I have no doubt that highly evolved spirit beings have a vast knowledge about all the events on earth, concerning past, present and future. They can tell us why storms, major fires, wars, droughts, floods, earthquakes, and social unrest happen, how we humans contribute to them and how we can prevent them or reduce their impact. We have largely ignored spirit beings of all kinds, thinking that because they were invisible, they were also insignificant. It seems that the time has come when we can, will and must communicate with them. They are willing to help us clean up the mess we have made for ourselves with pollution, plastics, climate change, water scarcity, soil depletion and so on. It is certainly also a matter of survival. Respecting our free will, they are waiting patiently until we are ready to cooperate with them.

In this book I present communications that I have experienced with spirit beings of all kinds. Some are recent conversations and others are already published in my earlier books. They range from the simple everyday concerns of these beings, to the more sophisti-cated ones I have added towards the end of the book. They show how precise and immense the knowledge and insights of these beings are. We need to build a rigorous culture of communication with them that enables us to know who is 'on the phone', how to question and understand them, and how to learn to distinguish valuable from non-valuable dialogues and partners. Can we avoid finding ourselves in our own Disneyland of projections? It entails keeping our fears, feelings of inferiority or superiority out of the way. Can we learn to speak from the depths of our heart, from our higher spiritual self? This requires a high degree of sincerity, which is absolutely necessary if we want to make communication credible and above all useful.

Metaphor

Imagine that we are living in an apartment building or a housing estate that functions well. The postman comes regularly, there are contract services for plumbing, electricity, elevators, cleaning, maintenance of flower beds, rubbish disposal, small repairs, watering of trees and plants, everything you can imagine is done while we are at work, travelling or asleep. Amazing!

And yet, until today, it has not occurred to us to greet these invisible service providers and thank them for their work and care.

This is exactly our attitude towards all nature spirits and invisible light beings that take care of nature and much more. We could start by saying hello to them and showing our gratitude or even asking if they are well and what we can contribute.

All space around us is filled with consciousness, whether we are aware of it or not. Every part of the invisible is consciousness and is always inhabited by a being: landscapes, trees, every object, every thought, are linked to beings. Consciousness is always related to a being.

Behind all manifestation there is a consciousness, that is a sentient, intelligent being. [11, 12]

"All that you see, all that you touch, hear, feel, know is consciousness/awareness made visible." Letter 5 / p.16 [11]

Just as a human being is not just skin, bones and brain, the physical tree itself, the flower or the landscape are not just a 'tree being' or a 'flower being'. We need to recognise the sentient, feeling consciousness behind them and the myriad of beings contributing to their wellbeing.

Any newly created space – physical or non-physical - fills immediately with a new being. A thought creates a circular energy whirl in the etheric, thus creating a new 'sub'-space, immediately occupied by a new created being. Similarly, a seed becoming a shoot, creates a space immediately taken up by a new being.

So, what is a being, what is their function? And why is consciousness necessarily always linked to a being?

All that happens on a subtle energy level. Subtle energy is all phenomena that are not grossly material and are beyond the electricity known by conventional physics. Subtle energy is normally invisible to our physical eyes, though clairvoyants, sensitive hands and radiesthesia tools (like a Hartmann-Antenna) can pick it up. Marie Curie, twice Nobel Prize winner, is said to have used radiesthesia. Usually nuclear and quantum energy are not considered part of subtle energy, although they are overlapping at times.

Using a Hartmann antenna

If in quantum physics we can accept the phenomenon of quantum entanglement, a force we cannot yet fully understand, perhaps we can also accept that there are comparable forces in radiesthesia that cause a tool like the Hartmann antenna to indicate YES and NO responses.

Communication between humans and nature spirits or beings of the divine field functions with the help of a translator being. They are Sphere 7 beings (see p. 77). To enable communication with invisible beings, we have to connect with our feelings and then formulate our question. The question is translated by these sphere-7 beings. These invisible beings then answer with their kind of feelings, which in turn are translated. The interface with us human being is our upper reflector ether. [2]

*) Quantum entanglement is a physical phenomenon that occurs when a group of particles is created, interacts or is in spatial proximity such that the quantum state of each particle in the group that cannot be described independently of the state of the others, even if the particles are separated by a great distance. The issue of quantum entanglement is at the heart of the differences between classical and quantum physics: entanglement is a key feature of quantum mechanics that is absent in classical mechanics. (Wikipedia)

Why is the invisible dimension invisible?
What is the use of dimensions being invisible?
Why can we not see e.g. nature spirits or our guardian angel?

Practical reasons for invisibility
The immense complexity of the invisible reality would make focussing and operating in the physical extremely difficult. Our primary task, in order to successfully fulfil our role in our 'Earth Game', is to stay focused on the physical, on handling and evolving our emotions, seeing the consequences of our thoughts in physical manifestation and through this developing our mind's potential. This includes proper grounding and a thorough incarnation. We could not function in a balanced way otherwise.

What I can perceive of the energy dimension, alone on the etheric level, is so complex, we would get rather disturbed to see all its structures and energy flows around us, while having to handle physical tasks like making a cup of coffee. A coffee machine e.g. has a machine spirit and a number of sub-spirits and elementals.

In addition to this etheric level, we would see all thoughts, atoms and emotions of persons present in the room. And these are again only some of the energy fields we are part of. In physical life as well, you don't see the inside of a motor or machine without Xray or suchlike. Except when taking it apart for instance for repairs and maintenance. Why should we see the intestines, blood vessels and all etheric energy lines of every living being when meeting them?

Possible philosophical reasons why we usually don't perceive the invisible dimension: We are spirit become physical. That means the essence of our being comes from the invisible world. Is there a 'game of life' on earth whose purpose might be, amongst other things, to work us through the illusion of physical and visible appearance so that we remember our invisible spiritual origins?

My conversation with the landscape angel of old Westbury estate gardens on Long Island.

The landscape angel has been unhappy since 1944, when the former owner probably died. His wife took over managing the estate gardens and went against her husband's principles by introducing trees from Canada that created conflicts with existing local trees in the park. The couple's children who now care for the estate, would understand the landscape angel, but do not know about that change of principles after 1944. The principles were based on a real understanding of local trees and how Canadian tree species would go against this local harmony. As the Canadian trees have been growing for over 78 years landscape angel says, there is nothing I could do to relieve his suffering. I am referring to about 80 leaf trees coming from a specific area in Canada, near Ottawa. The landscape angel said it would help to have a conversation with the new owners about that problem. Cutting down the Canadian trees is not required. The landscape angel appreciated our talk and feels better though the problem is waiting to be resolved.

Landscape Angel

My path started with university studies in economics and business administration. This brought me some notion of academic rigour and methods. Soon after graduation I began a long study of my inner world through therapy, personal development and twenty years studying spiritual healing and meditation with Bob Moore. The essence of his teaching was centred on exploring and experiencing feelings. I continued that work through teaching, musical improvisation and the feeling dimension of music, and exploring the mystical devotional path. Observing through my personal and professional work the trickiness of the ego, I came to believe that real long term progress can only be achieved by grounding one's path on four independent pillars of development: objective energy awareness exercises, authentic spontaneous artistic expression, a devotional practice (basically believing and exploring the divine dimension), and meditation. Using four such pillars prevents us from getting trapped in the illusions of a single one of these paths. Meditation can lead us to avoid important parts of the inner work with our emotions, as can prayer, artistic expression or even energy awareness exercises alone. If you only walk one or two of these paths, you may not sufficiently recognise your own blind spots. Our ego structures require a greater effort to be recognised and overcome.

The exploration of the invisible, including contacts with the beings of the invisible dimension, grows through this inner transformation work, as it gradually gives us access to the upper layer of the reflector (warmth) ether* and through it to universal wisdom and nature spirits. * see page 98

I am also very grateful for having discovered the *Christ Letters* [11], channelled and published in the beginning of the 21st century. These letters give us an invaluable insight into the laws of the Universe, the laws of Creation and meet the insights coming from the quantum field theory.

A dialogue with the Loire River Elemental
(June 10th 2021)

A while ago you showed me where you had your permanent anchoring place on the Loire River System, near Bonny-sur-Loire.
Is that location still correct? - Yes
Are you a large water elemental? - Yes
I aim to introduce you to the people working for the Rights of Nature and especially those concerned with the Loire valley, so it would be useful to help us understand what role you could play in this process. I do not have a special mandate to do so at the moment so is it ok to interview you now? - Yes
How to proceed? Do you have a particular take on the subject? - No
Let me start with my usual opening questions. Are you well? - Yes. Can I do anything for you at the moment? - No
Would you be interested in a co-operation with people working for the rights of the Loire River System? - Yes
The Loire River Systems includes all affluents from their source to the Loire River estuary? - Yes
Is there a particular section that needs to be looked at today? - No
Is there a particular theme that needs to be looked at today? - Yes
Can we talk about that? - Yes
Does it concern pollution? - Yes
Is it pollution caused by humans? - Yes
By industrial pollution? - No
By used waters? - No

An elemental is a nature spirit linked to one of the four elements: earth (*gnomes*), water (*undines*), fire (*salamander*), air (*sylphs*). G*nomes* (earth elemental) can go from tiny ones to very large earth elementals.

Through the writings of Federico Faggin on consciousness and quantum field theory I found many parallels to the *Christ letters*, sensing a possible meeting with the spiritual science of the invisible. Faggin made me realise how the different energy fields that we humans are part and consist of, each have a completely separate and individual language. One field of experience does not necessarily have a direct contact with or knowledge of other fields. As if they all were quite independent worlds of experience. [12]

The game of life on Earth
Human life on earth admittedly has tough rules that only make sense when we see it as an inner growth process. Everything we think, say and do has consequences that sooner or later come back to us like a boomerang. This seems to be an essential part of the game: Through the effects of painful emotions or pleasing good feelings, we learn what the consequences of our actions feel like. From this we can grow and prioritise spiritual values in our choices. This in itself is a remarkable invention of the creators. Further on, I discuss why we often need adversity to learn.

There is much evidence that even before conception we set the broad lines of our own life goals, what circumstances we want to be born into and what challenges we want to overcome in this life.[10] The challenges are not the problem itself. It is our abilities to overcome them, our beliefs and our spiritual faith that need to grow. The 'rules of the game' are the framework and not the problem. The problems that arise are linked to our inertia, intransigence and lack of understanding of the purpose of life.

The goal seems to be an ever-expanding evolution, which in recent centuries has increasingly focused on the growth of mind. We are learning to go beyond the lower intellectual mental, which is strongly influenced by our emotions, and gain increased access to the upper mental potential, operating beyond the ego. The ego is not the problem but the hurdles we need to grow spiritually by. It seems that there can be no lasting personal transformation except through the help of our higher self, the spiritual.

Tourism? - No
Fishing? - No
Boats? - Yes
Any other source or type of pollution right now? - No
Are we talking about barges transporting goods? - No
About individual boats? No
Located near your estuary? - Yes
Near Saint Nazaire? - Yes
Are we talking about the 'Port de Comberge' at Saint- Michel-Chef-Chef? - Yes
Obviously for you your estuary up to the Pointe de St. Gildas is part of your River system? - Yes
The pollution caused by the harbour and the private boats is what you are pointing at? - Yes
Is that all you wanted to say today? - Yes
Did I understand your answers correctly? - Yes
Thank you. Your indications are unexpected for me and seem completely useful.

As usual, this interview shows coherence, consistence, unexpected aspects, and is useful, and precise. We had no particular agenda or mandate today and both would influence the outcome of a communication. Although I receive the answers in a Yes/No form with the help of a radiesthesia device, a small Hartmann Antenna, I sense that there already is a non-physical being helping to formulate the questions. The questions are thus almost part of the answer. The interviews always feel much like having an intelligent 'human' on the phone: grounded, direct, practical, and non-egoistic.

One consequence of not seeing the invisible has been, that we manipulate the physical world without understanding the manual, the laws that are governing it. This led us into the impasses we are experiencing now. It is as if we were shown how disastrous this superficial handling would end up. All our environmental problems force us to get a better understanding of the laws at work. The holding on to a solely physical-material, spiritless worldview is to prove to be an illusion. If we seek to restore harmony in the physical world we need to become conscious of the invisible laws and how they are operating?

As far as I understand it, the invisible consists of numerous planes and dimensions. It takes some skills to tune into one or the other and not all at the same time. Comparable to a radio set, you would not want to hear all the radio stations at the same time. Accessing these different worlds implies knowing how to tune into one and handle its language and its specificities. As shown by the reality of very small elementals, gnomes (earth elementals) and undines (water elementals) don't perceive nor communicate directly with sylphs (air elementals) or salamanders (fire elementals). They live in separate layers of the etheric, thus in quite different 'worlds'. It is difficult for us to imagine that different energy fields are in actual fact quite different 'worlds'. Beings of one plane, just like us, don't necessarily see all the other planes.

The perception of the invisible is not just an on/off event, seeing or not seeing. There are a number of ways to perceive, and all these ways are gradually and individually developed. Besides our physical eyes, we have the 'third eye' and the 'eyes of the etheric'. There are more that twenty different types of clairvoyance, and that means many ways of seeing only partial aspects of reality.

Many of you want to be in a process of opening your heart more deeply. This takes great courage because it requires your willingness to let go of everything that is not you. When you let go of that, you become more authentic. This is a process of transformation - of healing inner wounds. And we can help you with that - we can help you. You have to understand that we like to see you become your true self, because in that way you will be able to come into deeper real circulation with us, and we can share the experience of divinity, which is the deepest reality there is.... And then you can bring this into the world.... You see, that's the whole idea... to pass it on.

Ignatius of Loyola
Channelled by Eva Høffding

I witnessed Ignatius speak directly to me or to another group member many times. I was there when Eva channelled this message from Ignatius, as well as messages from Pegasus and the Black Virgin. It is a beautiful experience to feel that someone is really here, even if they are only visible to some of us. Ignatius has his own poetic language. This makes him distinguishable from other spirit beings quoted in this book here.

Eva also channels our former teacher Bob Moore, in his distinctive way of speaking. It was invaluable for me to feel the reality in these presences and communications. It made it simply believable that there are intelligent real beings communicating with us.

We also have our feelings or e.g. our faculty of clairaudience that play a part in the perception of subtle planes.

The visual aspect is not the only one. The contact with beings of high levels can be overwhelmingly powerful. Seeing gets us involved in an encounter, and thus becomes a total experience. Seeing is knowing and getting involved. On that level the contact over yes/no answers may sometimes be more appropriate or the only one available.

What is the use of perceiving and understanding the invisible?
The beings of the invisible dimension have immense knowledge, wisdom and experience when it comes to the functioning of nature. How to keep nature in balance, how to understand the consequen-ces of human action or non-action? Human beings are now experiencing the effects of ignoring the wisdom of these invisible beings for centuries. They may not have the answers to everything but certainly wish to contribute to finding solutions for new problems as well.

In the many communications I had with beings of all kind and levels, they always are precise, consistent and sometimes come up with concerns we had not thought of. So, the newness of what they contribute is also a valuable aspect. Communicating helps to eliminate misunderstandings and unnecessary fears. It brings respect and caring.

Obviously, higher beings of the divine field have the capacity to inspire us and to share the deeper meaning of what is happening and the purpose of our lives. We first always need to check our motivation why we at all would want to communicate with or perceive invisible beings

Angel of the Vézère Valley
(April 2022)

Walking the dog today, I passed the spot
where the Angel of this part of the Vézère
Valley has its anchorage, near St Leon sur
Vézère (Dordogne, south-west France).
Nearby, I met an elderly man who was about
to pitch his tent. I showed him where the
angel was, not far from where he was
standing.
"What is his mission?" he asked.
So I asked the angel, "Are you all right
today?" - no
"Is there anything I can do for you? - yes
"Do you want to tell us what the problem is?"
- yes
"Is it to do with pollution? - no
"With the blockage of the riverbed?" - yes
"Near here?" - yes
"Upstream?" - no
"Downstream?" - yes
"Less than 1 km?" - yes
After a series of questions about the exact
distance, he answers in the affirmative
470 m.
"Is it an obstruction due to the accumulation
of earth, debris and driftwood?" - yes
I went and he showed me the exact spot.
Sure enough, there were several small trees
in the riverbed on the other side.

Fear of the unknown

Part of the question concerns our fear of the unknown and of the invisible dimension. Yet, why should we feel fear about discovering reality? What is the function of fear?

In the human energy system, the emotion of fear is centred in the stomach area, energy wise called the solar plexus chakra. Without going into details about that chakra, we know that the overcoming of fear happens through a better understanding and the attitude of love. That chakra is influenced by the element of fire and by the astral. Emotions from outside affect us mainly through the solar plexus. When we don't feel strong, we feel fear of being invaded by anything unknown. Fire is linked to light and spirit and understanding.

Isaac Newton and overcoming the confusion of the Middle Ages*

Since the 17th century Newtonian science moved away from recognising any conscious intelligence in nature. Isaac Newton laid the foundations for classical mechanics. He also contributed to the progress of optics, astrology, and mathematical analysis. What is less known is that he had dedicated much time to theology and Alchemy. John Maynard Keynes, the eminent economist of the 20th century, considered that Newton was the last 'magician' and 'alchemist' of the Western world. [9] Newton was thus familiar with both aspects: the exterior material world and the invisible dimension explored to some extent in Alchemy. My thesis is that laying the solid bases of physics of the visible physical world was fundamental for our civilisation and our science. These 'solid' foundations had become an absolute necessity in order for science and the world to move away from the superstition, confusion, fears, and obscurantisms of the Middle Ages. There was great confusion about the role of the invisible, how to perceive it and how to understand it in a non-egoistic way.   (from my book [21])

"Can you and your colleagues take care of this yourselves?" - yes
"Does the area you are responsible for go from Terrasson to further than Bergerac?" - yes
"Is your job as an angel different from that of a great water elemental who also watches over the river?" - yes
"Simply put, angels like you bring an element of inspiration, while water elementals deal with the daily 'technical' problems?" - yes
"So, it's not you at all who 'rolls up your sleeves' to clear the piled-up material?" - no
"But it is part of your daily concerns?" - yes
"Also, you wanted to show us a practical part of your duties, since the man had wondered about it?" - yes
"The problem will be fixed by the coming rains with the active participation of the water elementals?" - yes
"Thank you very much for this conversation" - you are welcome!

This conversation illustrates that communication with nature spirits is not always about sensational information. The same could be said when talking to a farmer, a bricklayer, a baker or a person in an office. They do a lot of everyday tasks that have to be done properly but are not necessarily extraordinary.

Although this brought about a fabulous progress in analytical thinking, allowing the material part of nature to be dissected and analysed in great details, it led us to lose the meaning of how and why all was connected. This very thinking divorced us from the wisdom of a native connection with nature …and often made us destroy their ecosystems and ours.

The Western world has taken over three hundred years to move away from the confusion regarding invisible phenomena and then to learn how to discern. It is only in the 20th century with the discoveries of the theory of relativity, subatomic particles, and quantum physics that science has become ready to explore the invisible dimensions any further.

The fact that the intelligence of nature and of its invisible beings was pushed into a corner and mostly survived as a memory in fairy tales becomes understandable. The scientific world will only fully recover from being stuck in materialism once we will have integrated the discoveries of quantum physics.

A change of paradigms is under way. 2000 years ago, all the cultures in Europe: Ancient Greece, Rome, Celts, Nordic countries had for example, gods of thunder, of a mountain, and spirits of the wind. These names or concepts corresponded to the understanding of those times.

Ignoring their existence for centuries, we find ourselves in a position of hardly knowing what the possible function and fields of competence of different kinds of nature spirits are. We believe that nature seems to function pretty much without us, but does it? does it function well? What benefits would a conscious partnership bring? The peculiar thing, about the dilemma of material physics in an ultimately invisible world of energy, is that nobody can prove the existence or non-existence of the invisible.

4 Spring Water Nymphs
(April 2022)

Recently I went for a walk with our two dogs in a
small valley. Only in spring and after long
periods of rain does the water of a small spring
flow over the path. When I arrived home, I saw
four pillars of light standing near the spring as
an inner vision. The next time I went there, I
perceived the four pillars and asked them, as I
normally would:
Who are you? They said: Water nymphs.
Are you all right? - No
Is there anything I can do for you - No
Is there something you want to tell me - Yes
They said there was a disturbing energy nearby.

Western science may be proud of researching what part of the brain is affected by moving the left little finger, and many fascinating aspects like this. Scientists do this because they believe any discovery has to be proven by its numerous identical repetitions and only then would they feel taken seriously. That is their particular procedure with specific codes of behaviour.

Yet, as soon as you try to find out who an interesting actor, stateswoman or inventor really is, you don't get very far by sending persons to interview them. Each one, if they really ask interesting deeper questions, will get a personal meeting, a personal impression, but comparing the different interviews will not lead to a 'scientific' answer about who that person is, what they think, believe, stand for. You may trick yourself with a zebra, by getting observations that match another scientist's observation of the same zebra. But they will not be able to agree on deeper aspects of whether that zebra feels, has a group soul, and has a mental activity or only an instinctive intelligence.

The science of the invisible dimension
Exploring the invisible dimension depends entirely on personal experience. While eastern science has a long experience using this approach in a perfectly rigorous and methodical approach, western science went along the path of collective experience. Any research, following that approach, has to be unpersonal and verifiable by third persons. This western approach has produced a science that limited itself to the material, the physical, measurable world. Yet, not dealing with the personal inner dimension did not produce tools that dealt with the inner world of feelings and emotions. It is only through the simultaneous exploration of quantum physics and the psychological techniques that the west is slowly opening itself up to go beyond only the physical.

I found another being next to the water nymphs. It turned out to be **a fire elemental**. It was upset. So, I asked it if there was anything I could do for it. It said:
- Yes
Do you want to tell me what it's about? - No
Do you want me to send you distant healing with my harp when I get home? - Yes
I did, and shortly afterwards I received the news that he was now well.
As I walked away from the source, I was flooded with a wave of sadness. Obviously, this fire elemental had experienced a deep tragedy.

As I am writing this paragraph, I asked him if he wanted to tell me now what his feelings had been about the other day - Yes

After a series of questions, answered by yes or no as usual, he said that he and his colleagues saw the coming war in Ukraine and the excessive misuse of firearms that would cause so much suffering. He and his fellow fire spirits felt it was a personal, overwhelming tragedy that fire would be misused to cause such suffering.

First nations have maintained the contact and belief in the existence of the invisible. While keeping a holistic, compassionate sense of and deep respect for the meaning of life, visible and invisible, they could mostly not go beyond a mythical belief system. The upcoming analytical western approach brought methods of exploration that allowed a progression into more details, leading often into excessive dissecting and losing the purpose of life. This analytical approach widened the use of our mental capacities, in a first period particularly the intellectual, lower mental aspect, though often leading to the trap of losing oneself in books that talked about other people's experience and neglecting one's own. The further development of upper mental part requires the overcoming of the ego.

Many explorers of the spiritual science, concerning the invisible dimensions, have through the centuries contributed a wealth of insights and paths to explore, like Rudolf Steiner, Edgar Cayce, Harry Edwards, Alice Bailey, Sufi masters, my teacher Bob Moore* and many others. However, their suggestions about accessing personal experience of the inner world were often not followed rigorously enough, as this takes years, if not decades of dedicated work on approaching one's inner world. This can be done through dream interpretation, meditation, energy awareness exercises, but always would have to be grounded in a daily discipline of handling one's physical reality of relating to others, to work, money, emotions and the love of the physical world in general: pet animals, children, family, nature.   *) spiritual teacher, Denmark, 1929-2008

Exploring the invisible cannot be done through solely believing in external authorities. The only real way is to walk an individual path of personal experience, strengthening one's own authority and learning to go clearly beyond the illusions of the ego structures.

In November 2017, I made contact with the **elemental being of the River Nore in Ireland**.

The elemental is located at its northernmost bend near Bishopswood, where the river begins its journey southwards. The River Nore (Irish: An Fheoir) is one of the most important rivers in the south-eastern region of Ireland. The 140 km long river drains approximately 2 530 square kilometres in Leinster and Munster. The river is home to the only known population of the endangered Nore freshwater pearl mussel and much of its length is designated as a Special Protection Area. According to the river's water elemental, for two years the Abbeyleix Water Authority had been discharging too many chemicals into the sewage system. Many of the chemicals eventually accumulated in the Nore River. These water elementals are responsible for all water systems (lakes, ponds, septic tanks, canals, etc.) in their area, not just the river.

Parallel Worlds

The different aspects of feelings allow us to distinguish completely different parallel worlds, where, should one exclusively be living in one, one would not know anything about the other worlds existing simultaneously. The world of the mental dimension e.g. consists of thoughts and pictures and differs from the astral peaceful feelings and painful emotions. The sensations in the physical body do not have thoughts or emotions and live on a totally different register of experience; however, they provoke thoughts and feelings and are very much linked to the etheric storage. The distinction within the spiritual dimensions again is possible through higher and finer feelings. Even on an etheric level, the difference between the chemical ether, light ether and reflector ether is so vast, that gnomes* and undines, working and existing in the chemical ether don't know the existence of the salamander operating in the warmth/reflector ether. They cannot communicate with each other except through the help of the Devas. *) see definitions on page 79

Four Levels of feelings

Recent research in philosophy suggests that consciousness is essentially based on feelings. This means that it is not necessary to have a brain in order to have and explore consciousness. We need to distinguish different levels of feelings:

1) **bodily sensations**
2) **painful emotions** - anger, fear etc (lower astral)
3) **uplifting feelings** - joy, compassion, serenity, (upper astral).
4) **higher spiritual feelings** allowing us to compare and be aware of two or more distinct phenomena in a field, we hardly have words to describe. These finer feelings can additionally be experienced as shapes and forms, colour, texture, movement, expansions, and many more. However, these feelings are clearly distinct from the first three categories. The discernment between

On several occasions the **water elementals of the entire river systems** of the Dordogne, Lot and Loire rivers contacted me.

Abundant rainfall had brought unusual amounts of earth into their rivers and they asked me to send distant healing to help them to cope with it. I later understood that this helped them make connection with beings of the Divine Field, who they seem unable to contact directly. Afterwards they reported that the distant healing had worked. The numerous contacts and explanations I received over the years have made me aware of the many daily tasks that elementals deal with, that we are usually unaware of.

The fact that dialogues with several water elementals of river systems appear in this book is due to the worldwide success of the Rights of Nature movement* in granting rivers legal and sometimes even constitutional rights. It reflects a growing awareness that water is the carrier of life.

*) I am a member of GARN (Global Alliance for the Rights of Nature) since 2019

these different levels of feelings avoid mixing them up and distorting the understanding of our perception.

I have been sending distant healing for many years. Distant healing is usually directed towards humans who have asked for it. As mentioned before, some years ago, nature spirits began to contact me and requested healing, telling me that prayer would be helpful. Afterwards they let me know that this was what they had needed. I wondered what was happening and what we actually can contribute that they could not do themselves?

What is Distant Healing see page 72

Conditions for observation and personal experiencing
Research into the invisible worlds and our inner feelings, requires us to be anchored in physical reality, able to deal with the physicality of our life's circumstances. Otherwise, we are faced with the projections of our inner fears, and unable to discern friends from foes. How can we reach a real discernment? How do we know who we are communicating with or where a channelling or suchlike comes from? Is it ego, is it 100% pure, only 90% pure, or less?

Discernment arises from proper grounding, from being conscious of our inner fears and of how we have learned to cut off our understanding from our perceptions.

Any communication with invisible beings has to be consistent and coherent (not contradicting itself over time). It needs to bring forward new aspects and be useful not just for us, but also for a particular area, if not the whole planet. The communication should respect our free will, and not have any ego aspect in it, such as wanting power, or to be put on a pedestal, flattering the recipient etc. It is always a challenge to ask clear questions and understand the answers properly. This takes some practice.

My contact with the **Pegasus beings** began in February 2013 with a dream of spaceships landing near the meditation centre I was visiting for a retreat. The next morning, after I had told the dream in the group, they unexpectedly transmitted a channelling through my friend Eva Høffding. What strikes me about the following text is its particular poetic language. Eva channels different beings and each has its own unique language.

We come from far away

We join your task force.
We see with mild eyes what is going on.
We prepare ourselves for a destiny with love.
Behold, we come from far away.
We come through your dreams
as a means of communication.
It is not always easy for us, but we see your
aspirations that we wish to honour
and wish to bring a response.
Our world is different from yours
and yet similar in some aspects.
We come from far away:
Below us - the earth, above - the sky.
We greet you with compassion, love and warmth in
our hearts.
We see you thriving in the time to come.
We will come closer to your reality.

They showed themselves immediately afterwards in a meditation with the symbol of a white winged horse. I remembered later that the winged horse is also called 'Pegasus' and that Pegasus is a constellation of stars near Andromeda. Afterwards I realised that there were a number of spaceships stationed near this centre, and also in the neighbourhood where we live in France.

Contacting invisible beings

A way to begin connecting to nature spirits is to care for them, by appreciating the beauty of a tree, a flower, a forest, a meadow, and thanking the beings involved in looking after these areas. We can greet them, and appreciate their contribution.

Perceiving denser energetic phenomena, such as fault lines or water currents in the subsoil, lower vibration earth grids (e.g. Hartmann and Curry grids) etc. can be developed through sensing.

There are a number of **preconditions for the perception** of more subtle phenomena with higher frequencies and these preconditions relate to personal transformation.

- Appropriate **discernment** is necessary and relies on good grounding (root chakra). (see my books on 'Chakras' and 'The Science of Spiritual Healing')
- **Accepting oneself** requires work on the hara chakra, among others. This avoids being influenced by with a sense of inferiority or superiority.
- **Overcoming the fear** of the unknown requires work on the solar plexus chakra.
- **Genuine interest** in invisible beings requires an open and compassionate heart.
- Finer perception implies a good functioning of the **third eye**.

All of this helps to activate the upper layer of the reflector ether, that gives access to nature, the universe and inner peace. (p. 98)

It goes without saying that patience, curiosity and open-mindedness are needed to encounter unknown dimensions. My path has led me through a training of 20 years with my teacher Bob Moore, a long practice of distant healing and meditation and a love for the majestic quality of solitary trees.

Some Principles of Co-operation with Spirit Beings
- The clarity and honesty of our motivation
- Our willingness to contribute to the common good
- Openness to divine impulses and inspirations

Consciousness is always connected to a being

We are not used to understanding 'Creation' as consisting of conscious beings of all kinds each inhabiting their own space: smoke beings, mist beings, fog beings, machine beings, sound beings, thought beings, cloud beings, curse beings, chair beings, scissor beings, nature spirits, wind spirits, animals, insects, trees, landscapes angels, river spirits, humans, house spirits, nation angels, or global spirits of feelings, to name only a few.

The Flensburger Hefte interviews with beings of all kinds and my own encounters with different beings, have gradually helped me to appreciate the wide phenomenon of 'beings'. I have experienced a number of what could be labelled as 'other worlds': ET's, beings of the mythological or magical realm. I described the magical realm beings in one of my books: the magician, the unicorn, the dragon, the light, the white fee, the candle bearing stag, the owl, the goblin.[7]

In the mythical realm there are numerous beings, usually present mainly in their geographic area of origin: the Nordic mythical beings, the Roman-Greek zone, the Celtic zone, African, North-America, Amazonas, etc. On one occasion I dialogued with a giant living in an important prehistoric cave with wall paintings in Dordogne. He had made his presence felt through a communication line at the crystal and complained about 'his' cave being used for tourism. See p. 86

Non-physical beings often contact me through energy lines that I sense converge onto a quartz crystal in my meditation room. I then proceed to ask questions about who it is and what it wants. I can also ask for these lines to appear around any object, including my body. The crystal is apparently not absolutely necessary for this.

But: What is common to beings of all kind?

Daniel Perret – Beings of the Invisible Dimensions
Right side: continuous explanatory text

Elemental beings in churches

On several occasions I came across the four elementals (a gnome, an undine, a salamander and a sylph) held into a position over circular mosaics on church floors (three in Liguria – Dolcedo, Imperia and San Paragorio in Noli) also in the basilica of eastern Brioude in France). It was as if the circular mosaic shapes had been deliberately placed to receive the 'forced presence' of the representatives of the nature spirits. It was clear that this had been done intentionally. A controlling astral being seemed to be functioning under a memory injunction of a human, sometimes from a long time ago. Its only mission was to maintain a pressure on the elementals, to stay there.

On this photo we have a circle with an air elemental stationed on it. He said that he accepted his task to stay there. This type of action in churches may date back to the influence of Saint Francis of Assisi. It hints at the importance of including nature and its sentient beings in our veneration of creation and the divine.

What actually is a being?

"A being can be defined as created by a **will** (of another being) that has an **individual existence** (material or immaterial), some kind of **consciousness**, a **memory**, a **purpose**, and **a will** – even when sometimes strictly limited – to stay alive and be able to pursue that purpose. These qualities give each being the **possibility to communicate** with other beings. Each being inhabits **a space** (a physical object or simply an energy space) that it is looking after.

While this means that all consciousness is always linked to a being and a space, this definition shows that there is a great variety of degrees of consciousness, more or less free will, potential of development or levels of expression. All beings have the capacity to accumulate memory, even if this happens, with some, rather like an automatic unconscious storage in their etheric of what they do and what happens 'around them' during their existence. They nevertheless keep to a minimum, for want of expression, a means of displaying or sharing the content of their memory." (quoted from [2])

Relating to actual beings, beyond the concepts of consciousness or space, brings something more familiar we can connect and know how to communicate with, even when the partner is a group being or a higher being and not just the single individual being. We often have a preconceived idea about the being we are contacting. These ideas may be a hindrance. We can create quite a projection about angels for instance. Although we may never know exactly what angels are like, we often diminish ourselves and put them on an unreal pedestal. They of course have an incommensurable wisdom and knowledge and live in a completely different world that we can hardly imagine. However in my experience they can be quite matter of fact, don't act superior and are helpful to cooperate with.

Interview with an elemental being of the 5th kind

I noticed a column of energy behind our house
and wanted to know what or who it was.

What type of being are you?
Are you a being - yes
Are you a light being – yes
(as opposite to a dark being)
Are you from the angelic hierarchy - no
Are you a nature spirit - yes
Are you one of the five types of elementals - yes
Are you an Earth elemental - no
Water elemental - no
Fire elemental - no
Air elemental - no
Are you a Type five elemental - yes
Are you a small elemental of that kind - no
Are you a large elemental of that kind - no
Are you a very large elemental - no
(there are only five of them in France at present)
Are you a medium size elemental - yes
(these are the four categories or sizes of
elementals)
So, you are a medium size type five elemental - yes
Are there at all any small elementals of the 5thkind
comparable to gnomes, undines, salamanders and
sylphs? - no
Your main task is to encourage the cooperation
between humans and nature spirits - yes
Is your location fixed - no
So, you are not attached to a plant, tree etc. - no
Can you move freely and decide where you want to
be - yes
Is your geographic area of action limited - yes
Is this area for you about one square km - yes
Yet, one does not find a being like you
every square km - no

"The universal awareness of being became the impulse of individua-
lised 'I'-awareness demanding self-expression. Life and 'I'-aware-
ness are synonymous in the dimension of matter. They became the
consciousness of 'matter'." Letter 5/ p. 21 [11]

"There is nothing in the universe that is not consciousness made
visible." Letter 5 / p. 17 [11]

'Visible' means here 'perceptible', whether visible or invisible to the
physical eye. 'Matter' meaning here all manifestation visible or
invisible.

"The nature of universal consciousness is intent, inactive and in
equilibrium...an infinite, eternal, limitless, boundless state of
powerful intent – pristine, pure, beautiful. This intent is to express
its nature." Letter 5 / S. 17 [11]

The Universal Purpose is to give individual form to Creation and
experience it. This is the reason why beings are created, even on
the smallest level of a thought, a musical note or any single stalk of
wheat or grass. All life comes in the form of consciousness manifes-
ted in a visible or invisible space and as a being.

"Consciousness does not exist in the particles, atoms, or molecules,
but in the (quantum) fields and the fields of fields of which particles,
atoms, and molecules are states. The fields are conscious, not the
states." Faggin [12]

This means that physical units like 'a tree', 'a flower' or 'a river' are
not conscious as such, but that the conscious being – the fields of
consciousness – connected to them are.

Conscious beings are essentially individual conscious fields of
energy that can have a physical body or not. Consciousness is
carried by a being and inhabits a space. Consciousness, being and
space are inextricably forming a whole. *Space* refers not to
emptiness but to a manifestation of life and consciousness on a
physical or subtle energy level.

Are there 15 beings of your kind in Dordogne (SW-France) at present - yes
Six of you have been for a while present on Eyssal farm [1] - yes
(I do not list all the questions that brought me to that figure and will show that procedure further down when explaining how I get a specific year)

The purpose of being here

Are you going to stay on this spot behind our house for a while - yes
Are you here because of my research into the cooperation with nature spirits - yes
Are you doing more than observing - yes
Are you bringing impulses to my work - no
Do only beings from the angel hierarchy bring impulses - yes
Are you helping to facilitate the communication between nature spirits and humans here - no
Do you mean 'not at all' - no
Do you mean 'not exactly' - yes
Do you explain to nature spirits what is going on in terms of cooperation - yes
Is this your main task here - yes
Do you have to motivate them to cooperate with humans - no
That is because they are by nature quite willing to cooperate anyhow - yes
Do you have to explain our way of working to them - yes
Is it because you yourself have an understanding of these processes of cooperation - yes
Have you been trained to understand this type of cooperation - no
You mean this is inherent to your nature as type five being - yes

When did they come into existence?

Are you a relatively recent creation, a new type of being - yes

Types of Beings

The types and numbers of beings in existence are incommensurable. An analytical approach can lead to discovering many more details and types of beings. The challenge of this approach is to be able to stay in contact with a sense of the whole. One type of being may belong at the same time to different characteristics. It would seem that new types of beings come into existence every day.

My aim in now including a brief overview of the different types of beings that exist is to help you to understand what beings are, what their function is and the diversity of beings that exist.

Sentient beings

Maybe it is best to think of most beings as being sentient: humans, animals, beings of the 21 levels, etc. 'Sentient' is defined as being 'responsive or conscious of sense impressions' (Merriam-Webster). Sentient beings are able to experience some kind of feelings. The term 'sentience' comes from the Latin term 'sentientem' meaning a feeling. Sentient beings can be physical or non-physical. In a strictly anthropocentric view, whether a being is considered to have a feeling, a sentient experience, depends on the human awareness (individual, group, state, law) that defines how that being is perceived. A sentient being can feel, perceive and sense to a varying degree. Sentient beings have an awareness of surroundings, sensations, and an ability to show responsiveness. Having senses (level 1 feelings) makes a being sentient, or able to smell, communicate, touch, see, or hear. The extent to which they have astral feelings like happiness, sadness, pain, joy and fear or higher spiritual feelings differs from species to species. Sentient beings generally have free will and an awareness of themselves.

Did you come into being before the year 2000 - yes
...before 1990 - no
...after 1990 - yes (This is a double check question)
...before 1995 - yes
...before 1994 - yes
...before 1993 - no
So, you came into being in 1993 - yes
Have all your type five beings been created the
same year - no
Have some been created before 1993 - no

Is your presence here mainly to reassure nature
spirits
- no
Does your task mainly consist of explaining to them
what is going on - yes
Have I missed out on an important task of yours -
yes
Are you here to learn more about this relatively new
way of cooperating - yes
Is the major reason why you are here at centre du
Vallon because of my healing work through the
crystal - yes
Did the Dagda* bring that type of work about - no
Did C bring that type of work into being - yes
Is it you, type 5 being, answering the last two
questions
- yes
Have we sufficiently described your work here - no
Is your work also bringing information up to larger
type 5 elementals - no
Is this because they know anyhow instantly what
you know - yes
Are you part of a work group with the other type 5
beings - yes
Is this work group studying this new type of
cooperation - yes
Have we sufficiently described your work here – no
*) Dagda see page 83

Beings with very limited sentience
While the individual beings of this category may be limited in their capacity to feel, it may be less so for its 'superior' or group entity. When communicating with them one would not always know whether it is the individual being or the group entity answering us.

There are non-sentient rudimentary forms of beings, such as machine beings, sound beings or thought beings, smoke beings, mist beings, curse beings, memory beings, control-beings, chair beings, scissor beings…

Every sound produced on a musical instrument and every thought creates an energy movement or structure that is immediately occupied by a being with a sometimes-limited life span.

The felt perceptions and feelings (category 1, 3, 4, page 29) do not create such separate beings as they are part of a natural, immediate cycle that connects us to All-there-is. The perceptions of a colour, sound, taste, smell, etc. are felt perceptions, as Faggin puts it. They may have no need to create a being and continue to exist as such, because they have no other purpose to fulfil than to be part of a circulation in the present. It is only when we produce a thought that is connected to a felt perception that a thought being is created.

A thought is created with a purpose: to be expressed, memorized, understood, trigger a reaction, or to create a circulation. Once this circulation is completed, the thought fully understood, let go of, it dissolves. The same applies to consciously created musical notes. Their being dissolves after a while when the note is heard and thus has fulfilled its purpose. As soon as they have fulfilled their purpose and are physically destroyed, for example in the case of a chair or tree, these beings dissolve or go somewhere else. It is the same with machine beings, etc.

Do you have to instruct and teach nature spirits how
to cooperate - no
Do you have to protect the process from intruding
beings - yes
Do you want to tell us about them - no
Have we sufficiently described your work here - yes
Do you want to tell us something else - yes
Are you originating geographically from a particular
place - no
Is it other type of beings that have created you - yes
Are they beings from the angelic hierarchies who
created you - yes
Are they beings from above sphere 10
From beyond sphere 15 - no
Do they belong to sphere 11 - no / sphere 12 - no
sphere 13 - yes
In this sphere we find beings called 'Kyriotetes' -
yes
Amongst the different types of Kyriotetes is it those
who have the mission to spread the teachings of
Christ - yes
Is this the reason why you are also called Christ-
Elementals - yes
This does though not refer to members of a
particular religion but to a quality - yes
Are there type 5 beings like you all over the world -
yes
This does not depend on whether one is Christian
or not
- no
Are new type 5 beings being created when a new
geographical area is ready to welcome them - yes
Does your work essentially consist of bringing love,
compassion, mutual respect and cooperation
between nature spirits and humans - yes
Is this the essence of your work - yes

Attempting an overview over the phenomenon of light beings

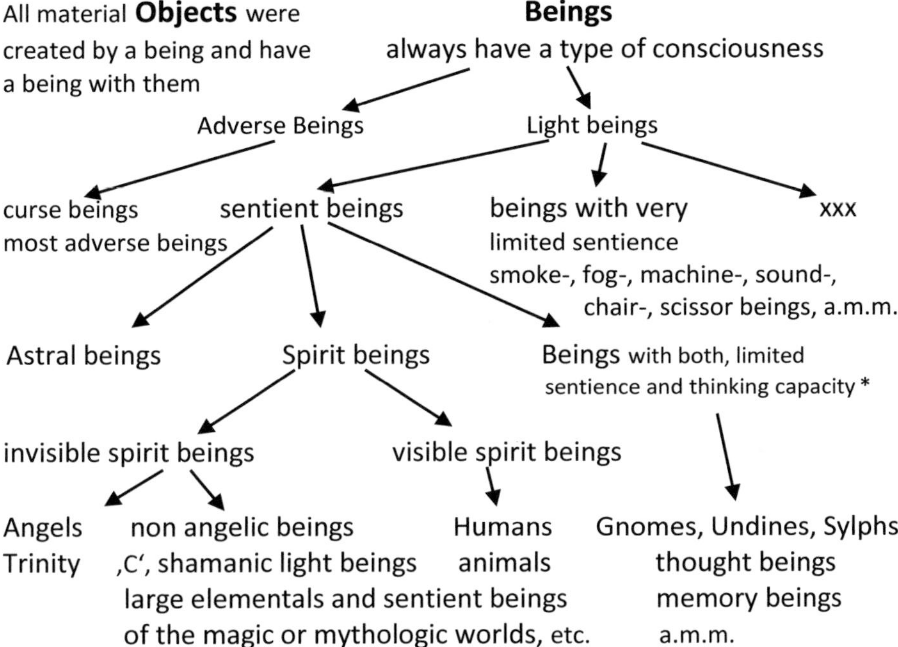

The limit of this scheme lies in the fact that concepts such as conscious-ness, mind, spirit, and sentience can have different definitions. The afore mentioned categories are thus probably not clear-cut and separate. Several criteria and categories may often apply to the same being.

*) These beings can communicate with us with the help of higher spirit beings; they don't have as such a fully formed spirit or ways of thinking.

Spirit beings versus 'non-mind' or 'non-spirit' beings
Spirit beings are non-corporeal sentient entities that possess a conscious connection to the spiritual dimension. Some people call this 'spirit'. Other people believe that any being that can express itself in an interview e.g. does have spirit, whereas spirit is not the thinking in itself but the content of the thinking. This opens the possibility of a 'non-mind' being connecting to the spiritual dimension through a representant like the group soul.

Can we move on to the question why this type of cooperation is so important for our time - yes

Is the reason that humans need to learn to become more responsible and aware of our relationship with nature - yes
To learn to love, respect and understand nature and all the processes that take place in nature - yes
So that we, among other things, avoid further damaging nature - yes
So that we also learn how to heal nature - yes
So that we become more active in restoring and maintaining harmonious coexistence - yes
This requires that people learn to ask the forces of love in nature (devas) for help, so that they fully cooperate in the processes of growth and healing - yes
So that we learn how our thoughts and feelings can invite these well-meaning forces to participate in these processes - yes
All this helps us to better understand the potential of our thoughts - yes
This will help us to have better harvests while using fewer chemical products - yes
Can we function without chemical products completely in our cooperation with nature - yes
For this cooperation to be a complete success and spread it will probably take some time - yes
We need time to learn and gain our own experience - yes
Can we do without this cooperation - no
Because it would bring more destruction to nature - yes
...more degradation of the soil - yes
...more toxins in our food - yes
...more diseases - yes
In the Dordogne we have relatively few creatures of the 5th type like you - yes
Is it because the learning process is so slow - yes

Yet, does formulating thoughts or have structured feelings mean the being has a mind, and is having a mind the same as having a mental energy field or aura?

There are many more spirit beings without a physical body than with one. And does having a mind or some mental capacities mean the being also has a conscious connection to the spiritual dimension? There is a vast number of spirit beings, whereof nature spirits and angelic beings of the Divine Field are only two kinds

Once they have fulfilled their purpose and in case of a chair or tree, are physically destroyed, these beings dissolve. All beings have the capacity to accumulate memory, even if this happens; with some beings this is rather like an automatic unconscious storage of what they do and what happens 'around them' during their existence.

According to my findings I suggest that a communication based on feelings is a direct, immediate circulation and quite different from an essentially thought based communication. A 'green thumb' gardener may communicate intuitively with Devas to find the right spot to plant a rose bush or young tree. No need for a verbal communication.

In the case communicated by the Loire Valley Elemental, a more elaborate dialogue was needed, involving words and thoughts (page 14). Yet, the feeling contact has to be there first to allow this type of dialogue to happen.

The unfoldment of Creation, both on a macroscopic and individual level, is built on two impulses: The impulse of **activity** and of **attraction – repulsion**. All impulses are directed by a purpose, in the case of Creation it is a Divine purpose.   see letter 5 / p. 24 and 25 [11]

Can you move outside your square kilometre - yes
Can you be in several places at the same time - no
Are there now more people who enter into cooperation
with nature spirit beings without consciously perceiving it
- yes
Is this made possible by people having open hearts - yes
Nevertheless, it is important that more people learn of
your existence and seek cooperation with you - yes
Is the explanation that many more people need to put
conscious cooperation into action - yes
Will a human will to that effect indirectly bring more
beings of the 5th kind into being - yes
At the same time, it is essential that this cooperation is
entirely based on free will - yes
Is increased cooperation urgent - yes
Is this really a race against the degradation of nature,
soils, food and quality of life - yes
All our sciences of biology need to be reoriented and
build on this cooperation and new knowledge - yes
This process will allow us to access the vast potential of
knowledge and experience that nature-spirit beings have -
yes
Interviews like this one will somehow make our contact
more comprehensible - yes
It is also important to show by means of comparative
photographs what happens to plants that grow under this
cooperation and others, as is done in Eyssal [1].
Would you like to add something else - yes
Are you happy if we publicize this interview - yes
Do you want to tell us something else - no
Thank you for this interview
Was it a pleasure for you - yes

The purpose of reporting these interviews is to show that, although invisible, we
are dealing with intelligent beings as if we were talking to someone on the phone.
Their invisibility should not keep us held in doubt.

Think tanks of spirit beings

My major steps in understanding the communication with invisible beings came through the contact to two think tanks* of spirit beings. One grouped around the field of sound, especially distant healing with the harp and similar multi-string instruments. The other being 'C', the collegium of spirit beings of Rocamadour [2] that I have been in contact with since 2012. I have researched with them on many subjects, which resulted in a number of books. The twelve permanent members of C are an interesting variety of beings: humans with past incarnations on earth, a Deva queen, and two teachers from outside our planet.

*) A spiritual think tank is a group of selected, not physically incarnated experts who deal with future-oriented innovations in a particular section or theme of the spiritual field.

Evolution and adversary forces

All beings I have described up to now are light beings, that is beings on an ascending journey towards the light or the universal intelligence. Many other beings are on a descending journey away from the light. We can call them adversary forces or beings. There is of course a question why they are here, why they are on this type of descending journey and with what purpose. Some accomplish tasks necessary for life, like the decay beings transforming a dead branch into humus. The polarity between light and dark is the reality of existence and is in constant movement. Light does not exist without darkness, and ascending beings do not evolve without beings on a descending path. There always needs to be a balance between the two. If you neglect the dark side, it will well up in your subconscious e.g. and if you neglect the light, the dark will take the upper hand.

In the Flensburger Heft (No 79 [13] p. 208) a Great Spirit (Etschewit) confirms what Wolfgang Weirauch asks: "...beings of ugliness … are necessary in order for a beautiful work of art to arise? Does a counterweight always have to be present when something is meant to come into being? … Etschewit: That's right. " continued page 97

Daniel Perret – Beings of the Invisible Dimensions
Right side: continuous explanatory text

Dialogue with C on opposing forces and the human ego

Do opposing forces exploit the weaknesses of
the human ego? - no
Is it weaknesses of the human ego that attract
certain adversaries? - no
(this is not what I expected).
Can we talk about the connection between ego
and adversarial forces? - yes
Does the human tendency towards fears
attract e.g. the Grey ET's? - no
Isn't there a connection between fear and the
Grey ET's that fuel fear? - no
Do you say 'no' because it is the weakness of the
human soul that attracts these forces? - yes
So, you are saying that the ego is neutral, so to
speak, like the natural obstacles on a walk in the
woods - and that it is the way the individual soul
accepts and deals with the challenge that can
create a point of attraction for these forces? - yes
So, it is the weakness of a person's soul force
that creates the problem, not the ego?
- Yes, exactly.
Can we leave this discussion as it is? - yes
Do we need to make some points clearer? - no
Thank you, that was indeed enlightening.
- It was our pleasure!

This is a good example of how spirit beings have their own way of seeing
that may surprise us.

We all know from personal experience, that we would often not move forward unless life gives us a push in the back. It seems to be part of a fundamental universal law that evolution needs adversity. Remember that light does not exist without darkness and I believe that adverse forces are obstacles that we need to learn to overcome. They are not the problem, but simply the hurdle, that incites us to use out strength and manifest it in our very personal way in this life.

The central practice to overcome obstacles is to keep in our hearts and minds the connection to the Divine or Universal Intelligence.

Celebration – Gratitude – Service

Celebration of the manifestation of the Universal Intelligence in nature and in all beings – filled with gratitude about being part of it – and in turn contributing the best we can to Creation.

The subject of adversary beings often brings up the projection of people's fears and negativity. Even mentioning the name of some of these adversary beings creates fear and confusion.

I look at these forces from the point of view of what they can teach us about our weaknesses and shortcomings that automatically create blind spots and become an attraction for such beings. I believe we are above all actors and not victims. Let us examine how we can grow through adversity by sorting them into five different areas of challenge.

1) Fundamental human rights
2) Christian values
3) Global imbalances and one-sidedness
4) Unbalanced human behaviour
5) Addictive behaviour

The teachings of the Black Madonna

"I am the Black Madonna. Blackness is my power. I work in all hidden corners so that everything can come to light. Motherhood holds deep silence. I am the mother. I am the Mother of humanity. I carry great silence in my womb. I am the starry night sky. I am the protective blackness and nurturing silence of the cave. From me springs all life. Take in the nourishment I give you, for it will bring you peace of mind and peace of heart. Merging with me is a journey. Let everything flow through you. Sink deep into your own black silence and then my love and my mystery will unfold in you.

I am the Mother of God. My beloved Son has changed the world. It was not an easy time for me to be with him on earth. My heart almost broke with grief at the loss of losing him, this beloved child I carried in my womb. But he was not mine. This was a service I could offer so that he could be brought into the world.

I was present when Eva Høffding received this message. I experienced a direct message from the Black Virgin in meditation, where she appeared as a face in front of me and said a sentence to me.   (continued page 54)

Fundamental human rights

Free will, creativity and spiritual growth are our birth rights. If we abandon these aspects we will sooner or later be seriously challenged. The particular beings challenging these aspects seem to come from the star formation of the Dragon, and are called Dracos beings. I have consistently met three types and their challenges should not be taken lightly as they do not let you get away with a half-hearted attempt to overcome their challenges.

- **Free will** - lizard-like Dracos
- **Creativity** – crocodile-like Dracos
- Right to **spiritual growth** – snake-like Dracos.

The challenge to our right to free will may insidiously creep into our life. We might for instance give up on a certain subject, thinking that it is too late or too difficult to change something. I have worked with such issues throughout my life, in the field of spontaneous, authentic musical expression, creativity and been aware of the obstacles that arise. Creativity includes also the way we think and operate in our daily life, not just artistic creativity. It is the very core of being alive and exploring new aspects.

Teachers like Christ have emphasised on a set of values that by evolutionary necessity became more predominant 2000 years ago. In order to separate his teachings from the loaded history of the Christian church, I call them Christ values and not Christian values. The above fundamental rights are part of it.

Christ Values

Faith – undermined by fallen angels for instance
Compassion – undermined by egoism; selfishness is an astral being very similar to a demon – and its addictive aspect.
Individuality – as opposed to group or tribal identity

54

What service do you want to offer to bring it into the world? I will be your mother, your teacher and your supporter so that your healing powers can be used for higher purposes. My breasts can fill humanity with divine nourishment in a never-ending stream of love. Where I am, there are no limits. There is only an endless giving, an endless presence and an endless love.

My silence is vast, is endless, is divine, holy. My silence touches everything and gives comfort to the suffering in an endless stream of compassion and love. This is my true nature, and this is why I was the Mother of Christ on earth.

If you let me work in your life, it will bring about a wonderful change for you. Apply me in your life because I can nourish you. I can nourish you with my divine milk. Spreading my words is of utmost importance, for these words nourish souls. I speak from my essence into the essence of humanity. These words must be heard, read and understood.

Please continue your good process so that your own divinity, your own glory can unfold."

Black Madonna
by Eva Høffding

Truth – confusion, mental-astral beings challenging it
Love – fear, ET Greys are often sitting in our auras when we persistently have problems with fears
Knowledge – Ignorance, backed up by an astral being, similar to a demon
Beauty - Contribution to beauty – challenged by ugly demon like beings
Celebration of life and Creation – undermined by boredom, depression, lack of spiritual beliefs and the astral/spiritual entities behind them.

Global imbalances and one-sidedness
Exaggerated, ungrounded ideals, cold beauty – **luciferian*** tendency
Denial of the spiritual, materialism, digital illusionary worlds – **ahrimanian*** tendency
Fanatical, destructive forces coming from beyond the zodiac – **Azuras***
Anti-Christ beings, opposing globally Christ's values, sorathic*
*) beings in the anthroposophical terminology

Human erroneous behaviour
Wandering, vengeful, or **clinging souls**. The souls of some deceased people can for selfish reasons cling on to their houses, possessions or obsessive emotions rather than leaving the physical dimension behind them and naturally evolve towards the light.

Black magic, arbitrary prohibitions, bans, curses, manipulation; often rooted in ancient traditions, a lack of respect of a person's free will by interfering can take many forms and sometimes can be carried out by 'professionals' who have learned such techniques. A tragic aspect of this practice is that often the people initiating an action do not realise that there will be a return effect of the

Interview with the Mother Earth Being

May I call you 'Mother Earth Being'? Yes. The intelligence behind nature is a complex phenomenon. It is impossible to do it justice here in the framework of a book. Yet, it is necessary to become as precise and as concrete as possible, so that we can appreciate and respect who we are communicating with. (The being is ok with this statement)
Can a dialogue with larger nature spirits in charge of areas contribute to any political decision process about the area? Yes.
Can this dialogue be expected to be like an interview with a human being? No.
Would the respective nature spirits in charge of an area or part of an area be expected to bring forth a dialogue partner representing them all? Yes.
When working towards the recognition of the rights of a whole river system or a wet land area, could the group of people working on the question count on finding a competent and representative partner amongst nature spirits? Yes.
Would that nature spirit be expected to make politically understandable contributions? No.
Do you mean 'not necessary'? Yes.
Does that mean the nature representative would come up with its own range of requirements and information? Yes.
So, their contribution could often be unexpected? Yes.
Would it always be a valuable and essential contribution to the political decision process? Yes.
When I use the term 'political' process can this include also possible legal procedures? Yes.
Can we expect that such a dialogue would attract other conscious beings that would help uplift the energy level of the process? Yes.

Boomerang they are sending out. They may not set a planned end to their interfering action which can result in them accumulating karmic effects sometimes over centuries.

Egregore: energy clusters of a group of human thoughts/emotions
Addictive behaviour – attracts demons; drugs, digital games, sex, alcohol, overeating, etc. used to avoid feeling and numb deep pain.
Elementals who were not respected or hurt by human action.
Consequences of **ignorant human behaviour**, like the excessive damming up of rivers that made it impossible for the river to regulate itself through its own wisdom (overflow areas for high water periods); or the negligence of periodically clearing the underwood that laid the ground for devastating forest fires.

The upcoming role of humans. In the interview with high spirit beings (Flensburger Hefte [13], page 178) it is being said that we are witnessing a time where some angels have started to withdraw from their role as coach for nature spirits, gradually leaving their place to humans. This is the result of an evolutionary impulse passed on by Throne angels and coming from 'higher up'.

My understanding of this is that humans need to learn much more about the functioning of nature, and that this can come about by us being given this task. The numerous contacts and requests for help I have been getting from nature spirits could be a result of this. I often do not understand how I can contribute anything but accept:

1) I play the role similar to a telephone switchboard operator of the forties, getting a request and making the connection to a required specialist in the divine field. Of course, I do not know who to contact, but this connection seems to be carried out by competent beings.

This rise in energy level would include more tolerance, understanding, compassion, patience, joy, and respect? Yes.

If a dialogue with the intelligence behind nature is not included in such a process, would that mean often having more difficult procedures? Yes.

More misunderstanding? Yes.

More ego battles involved? Yes.

Including nature spirits in the process would help diminish these issues? Yes.

Can one say, that including nature spirits in the process would automatically bring in 'spiritual energies', good will and the experience of higher beings and thus raise the whole level of consciousness and cooperation? Yes.

So, although the process might often bring in some unexpected angles and statements, the overall result would be significantly positive? Yes.

This means that the humans involved would usually need a greater tolerance towards an unusual reasoning? Yes.

Do you feel we should involve other aspects in this present interview? No.

Are you pleased with our dialogue? Yes.

Do you feel I have understood you point of view correctly? Yes. Do you feel you have participated already in the formulation of the questions? Yes.

Thank you for that dialogue.

I must say, I had not expected you to answer.

This would help to bring about more mutual understanding? Yes.

It would obviously bring in unexpected aspects of the natural area, helping to appreciate more its complexity? Yes.

Would it also help in raising more understanding of how the intelligence behind nature functions? Yes.

Addition 16.4.22: Are you identical with the Black Virgin and the Black Madonna? Yes.

2) I am often told, as part of the request for assistance, what the problems are, and thus learn about the problems nature spirits experience in their daily work.

3) I have learned that through distant healing, often recently with the harp and its sound clouds, that I can facilitate this connection. My experience with spiritual healing helped me to understand the importance of the upper reflector ether in this process of transmitting distant healing. [5]

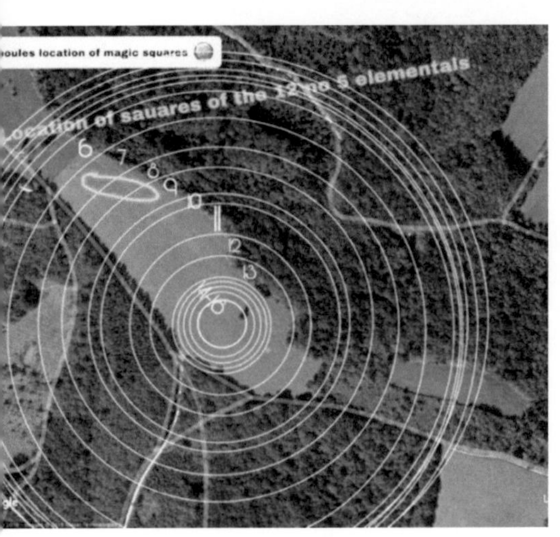

The 'assemblies' and magic squares as divine seals

About three years ago C told me that on Easter Sunday morning, at the sacred place near my home, an extraordinary assembly of beings from the whole southwest of France was going to take place. This assembly has met several times since, always on a Sunday morning, usually from 6 o'clock until 10 or 11. I would always be given the precise date and hours if I asked.

These assemblies seem to happen at the same local time all over the planet in specific sacred places, four for France, one in Switzerland, seven in Germany, etc. They would always gather 576 beings of all kinds from their geographic area. Why 576? Possibly because the three numbers add up to 18=1+8=9, 9 being the predominant number in the magic square and symbolising the heavenly circulation of energy (the upper circle) and the descent of that energy into the physical level, meaning divine inspiration.

12 March 2021: In my book 'The Intelligence Behind Nature' [2] I give a brief summary of the impressive process of reintroducing wolves to **Yellowstone Park** and its multi-layered consequences for restoring a more natural environment. A short conversation with the **'landscape angel'** responsible for the entire Yellowstone Park ecosystem revealed that he wanted readers to know that it exists and where it is located:
RJX2+X4 Devils Den, Wyoming, United States. –
(44.8498804, -110.3997381).

Each of these beings has their 'seat', a 'magic square' of 108x108 cm consisting of 3x3 squares of 36x36 cm.

The extraordinary fact is that these squares can be seen imprinted in the vegetation. The sacred place nearby is a field with grass where the squares can quite easily be detected visually and energy wise. These are elaborate squares established on the physical level by Throne angels, with a very complex set of information. The squares are there all year round, but only 'inhabited' during the assembly. In those sacred places the squares are placed on 18 concentric circles, starting at the periphery with circle No 1 and its nature spirits and landscape angel.

The beings of the outer circles (1-4) arrive already the evening before, while the inner circles fill up shortly before the beginning of the assembly, in the early morning. To my knowledge these squares on the field are the only known physical manifestation of the angel hierarchy.

I encouraged a friend to visit such an assembly place in the east of France, near Alaise, south of Besançon, and to look out for those squares. That particular place is located on the outskirts of this rural village with wide meadows. She photographed this square in the grass there.

How can it help to know about such squares and assemblies?

Obviously, the Throne angels wanted us to see these squares. The beings behind these assemblies also clearly wanted me to see, feel,

*On two occasions **Devas** of different sections of forest contacted me about a deterioration of their areas. Once it was because of too many badgers and another time because of too many stags. Contact with the group souls of these animals helped to rebalance the situation. The problem with the badgers had happened because too many offspring had remained in the section where their parents had settled. The youngsters needed to find their own territory.*

and witness these assemblies, because they always tell me which Sunday they will take place and in what time window. I can clearly measure the energy on the squares and in the field during the assembly hours. It is always significantly higher during the assembly than just a minute after its ending.

These assemblies, as far as I understand them, are taking place regularly to help nature spirits and other beings of the divine field to adapt to energy changes in the world. They need to have guidance, and to know who to contact for advice, when new types of problems occur. Not all changes are due to human activity. Some changes are brought about to help evolution.

I do not exactly know what happens during an assembly at these sacred places all over the world. It seems that a high Seraphim angel leads a type of meditation and silent harmonisation of all the 576 beings present. The Seraphim angel already knows what questions the participants are coming with. The activity of the assembly, I believe, happens on a silent and high level of blending. We can ourselves possibly get a sense of that quality during long silent meditation, where we have quietened the thinking process so that deeper levels of understanding can infuse our being. I can feel an immense joy spread over the field during the assembly.

Deducting from the structure of the 18 concentric rings, present in those assembly places, there would be 32 magic squares and beings per ring, with beings from the first 18 spheres of the divine field. C tells me there are 10 landscape angels on the first ring, 7 Dagdas, 11 Deva princesses and 4 tall Elves. I imagine they all are senior representatives of their region reporting back to their groups. Geographically they are evenly distributed over the territory of – here in the case of the sacred place near us – the south-west of France.

Two practical communications

The Deva of our forest

After installing solar panels on our roof, we had to cut down some of the tall trees. The deva was very helpful by showing us which ones we could cut down and which ones we could not.

She suggested that we leave the few very tall fir trees, because she has her anchor point at the base of the tallest one.

Replacing water filters

We use two special filter places for tap water. Both use refill cartridges, but of different brands.

These cartridges usually last much longer than the four weeks stated on the packaging.

There is virtually no way to find out when they have lost their effectiveness, unless you ask C. As always, the answer is quick, precise and unequivocal.

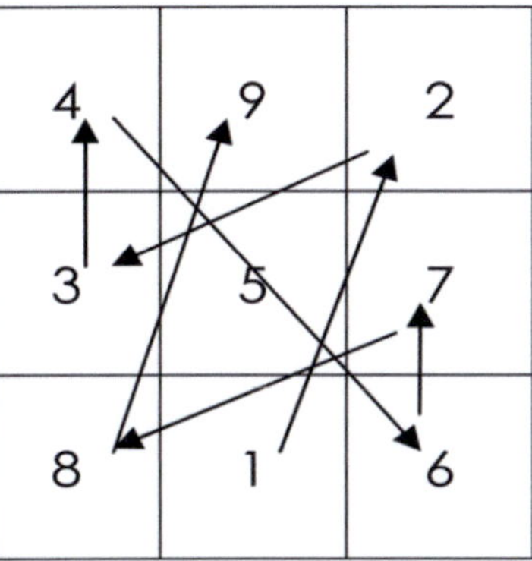

The magical or divine squares

Squares in nature are rare. I mention them here as they are mostly a near physical energy phenomenon, and when placed on grassy surfaces even show physically as squares with their nine sub-squares. They contain a large amount of divine information and were already known several thousand years ago in China as Lo Shu squares or Ba Gua. One peculiarity is the identical numbering sequence of the 9 inner squares. Each square has a spiritual symbolism forming a.o. the bases of the I Ching with the eight squares of the four sides. [4] The persistent sequence of numbering is amazing. It takes some concentration at first to see them in the grass, differences of vegetation occurring along the side lines. The energy within these squares is much higher than just next to them. I have taken dozens of photos of these grass squares. I even found two on the grassy patch outside the door of our village church: left side of church doors you have square No 1, left of the door is No 9 as I describe them in my book on the lost orientation of cathedrals. [2]

Same square 108x108 cm

Daniel Perret – Beings of the Invisible Dimensions
Right side: continuous explanatory text

I am moved by the
The song of blackbirds.
Since I have been interested in the
source of inspiration for my harp
improvisations for many years, I asked
C. if black birds are inspired by other
creatures? Yes
By an air elemental being? Yes,
From an air elemental being of a
medium category? Yes
And there is another source behind this
air elemental? Yes
An angel? Yes

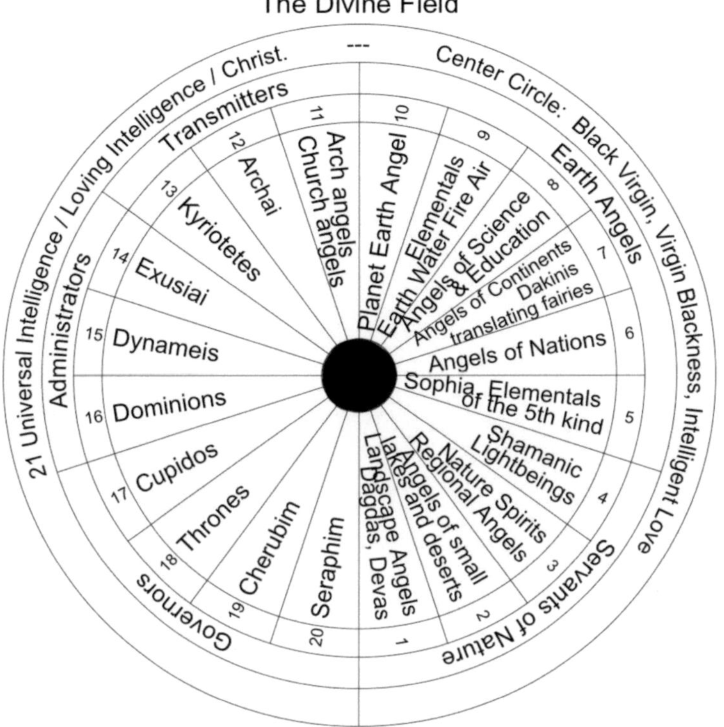

The Divine Field

I use this term to designate most categories of light beings, including the angel-hierarchies under the direction of Father Heaven or universal Intelligence. Nature spirits and affiliated beings we can consider as being under the direction of the Mother Earth Being. I continue to explore this divine field with the help of C and hope through this to evolve and deepen the understanding of its magnificence and beauty.

When I asked 'C' how many spheres or levels existed in the spiritual realm, that were source of inspiration for the arts for instance, they immediately answered 21. These are distinct from the astral or mental spheres of inspiration. [2] In a larger sense we could say all beings are part of the Divine Field, even basic forms like thought beings, machine beings, etc. as well as the descending and challenging forces. Nevertheless, I do not include these beings in

Daniel Perret – Beings of the Invisible Dimensions
Right side: continuous explanatory text

The machine being of our flat screen TV
C gave the following answers as to what the machine being of our TV consists of: our flat screen TV has one 'main machine being' and 17 'sub-beings':
- One dealing with the sound aspect
- One dealing with the visual components in general
- 6 colon beings: red, blue, yellow, green, orange, black
- One gnome dealing with the material aspect
- 4 fire elemental beings dealing with the complex electrical components
- 3 computer beings dealing with three different built-in computer programs/chips
- 1 coordinator sub-being that is not a computer program.

the term of 'Divine Field' here with its 21 spheres or levels.

We will never be able to fully grasp all the invisible beings of the divine field. Understanding the tasks and competences of the angels may remain an impossible undertaking, but in time our efforts bring us a certain understanding, above all respect, and thus open a door to cooperation.

Ever since first C communicated to me about the 21 spheres in the Divine Field, I have tried to understand what these spheres consist of and continue to explore this. Every time I meet a new being or learn more about a being, I add this to the list. My dialogue with spirit beings works exclusively through YES/NO answers. So, I had to address each being within a level and ask whether they are part of it.

At the beginning I drew on the nine angel categories that Dionysius the Areopagite, first bishop of Athens, had elaborated around the first century A.D. [14] He was an Athenian judge at the Areopagus Court in Athens. For the last 1800 years this scheme has apparently not been expanded on. He had indicated that his list was not complete and that the only one who would know how the divine field was organised would be God. C's answers were, as usual, very precise. They agreed on most of what Dionysius had suggested.

C guided me to add level 21, what we call the divine level, the Trinity level. C then insisted that we include a new level 17 for the Cupidos, the high angels of Art, Love and Beauty, which made immediate sense for me. The chubby little angel figures of our folklore with their bow and arrows of love, are just the popular, superficial version of these high angels of Art.

C was adamant that the Kyriotetes and the Dominions were not the same. This became another difference to Dionysius's

Interview with a group of shamanic light beings

that appeared at the crystal during the GARN Glasgow tribunal in November 2021

Shamanic light beings are nature spirits in the wider sense of the term.

In existence for thousands of years but seem to be being replaced by the elementals of the 5th kind . They belong to the time and ways of indigenous people: instinctive, compassionate, holistic, whereas the 5th type elementals blend in better with the evolution that has brought about a more analytical, inquisitive, explorative type. They communicate with animal group souls, insects, elementals above the very small ones (gnomes, undines, salamander and sylphs), Dagdas, Devas but not with angels, big Elves, beings of the magic or mythological realms. They were the direct invisible partners of traditional indigenous shamans.

scheme. At this point we had 11 categories or levels. The 10 levels below remained that I had to understand better. C agreed that nature spirits and landscape angels are part of these spheres. A more detailed description of all levels appears further down.

As part of understanding their different functions I grouped some levels together. Dionysius had used three categories of each three spheres of angels: the highest, the middle and the lower hierarchies. Having introduced two additional levels and not satisfied with Dionysius' designations – as it does not explain anything – I suggested the three first categories being the 'Governors', 'The Administrators' and the 'Transmitters'. I understand the 'Governors' as being the ones who set the directives, the main action lines. I believe in ancient times they were part of what people called God, as their injunctions were godlike. The 'Administrators' feel seem to be like the heads of administration, translating the directives of the Governors into practical directives. The 'Transmitters' are in charge of applying the directives in direct cooperation with the local angels and nature spirits of all kind.

Sphere 21 – Trinity, Christ, The Universal Intelligence or the Father principle and the Mother principle/Black Virgin/Earth Goddess

'The Governors'
Sphere 20 - Seraphim: high angels or spirits of light & fire, igniters, bringing divine inspiration
I find them in the aura around some harpists and their harps, indicating the harpists accomplished potential to send distant healing to humans and animals, beyond the more common capacity to send distant healing to nature spirits and nature in general (including landscape angels, Devas, Elves, etc.). This presence makes me feel that Seraphim are highly inspiring angels, very much bringing through the breath of the Holy Spirit.

Distant healing for invisible beings
How can this work?

A transformation of the relationship between human beings, nature spirits and beings of the angelic hierarchies is currently taking place. This is mentioned, among other things, in the interview series by Wolfgang Weirauch published in the Flensburger Hefte (13 p. 178) Since the beginning of the 1980s and until sometime after 2040, we are in a transitional period in which the angels are withdrawing to a certain extent and gradually leaving more tasks to humans as advisors to the nature beings. One effect of this is that humans have to learn much more about how nature spirits and invisible beings work in general and what their tasks are. In this way, humans will take on more conscious responsibility for nature and human actions. People will also gradually gain more understanding of how cooperation with non-physical beings works.

We can already contribute to this by sending distant healing, for example. I have the feeling that we function comparably to the telephone exchanges of the old times. We receive incoming calls from nature spirits, for example, and make connections with the angelic specialists. Then the communication between them works. This is also how I personally experience the requests I receive practically every day from invisible beings. Energetically, this works via our higher self and the upper layer of the reflector ether. [2]

I was aware of the presence of a Seraphim during the Sunday morning assemblies mentioned earlier. The energy of the Seraph is that of overwhelming joy, almost ecstasy that spread all over the field.

Sphere 19 - Cherubim: high angels or spirits of harmony and wisdom of God, Transmission of insight, knowledge

Sphere 18 - Thrones: high angels or spirits of divine will and life energy, giving impulses, structures and direction for humanity. I have found their influence at the source of fundamental spiritual structures like earth grids and magic squares. Once in a meditation I felt a vertical slab of granite like stone, a stele, thundering down beside me on my left side with an inscription on it: "The Path of Light is manifesting" and "The Power of Love is manifesting". I feel their energy to be very powerful, almost masculine and wilful, like laying the law down with authority.

Sphere 17 – Cupido: high spirits of Art, Beauty and Love
I find their energy around all beautiful works of art. The manifestation of beauty is extremely important, maybe especially in our time and certainly not just a hobby or a luxury. It is the visible expression of the Divine on earth. According to C Cupido came into existence in the 13th century.

'The Administrators'
Sphere 16 – 'Dominions/Dominations': they lead the earth angel as well as the angels of the continents and nations. They look after and guide people in their, sometimes contradictory, thought trends, spiritual concepts or ways of thinking about the spiritual dimension.

Sphere 15 - Dynameis – 'Virtues': directing the cycles of the planets; celestial harmony

74

Here is an excerpt from the Flensburger Hefte issue no. 79 [13], p. 102. The subject is of great topicality after all the fake news and organised lies around Trump, Corona conspiracies and Russian propaganda about the Ukraine war. The interview is conducted by the editor Wolfgang Weirauch himself. Etschewit is a high spirit being.

Wolfgang Weirauch: **What are demons?**

Etschewit: Demons are entities of the descending powers from the black realm, which are caused, among other things, by people's lies. They are also caused by people's false thoughts. Just as we are children of the angels, demons are children or spinoffs of the fallen angels.
W.W.: Does every lie that a human being speaks create a lying demon?
Etschewit: Yes.
And if it is a small lie, it is a little demon, if, on the other hand, it is a serious lie, it creates a full-blown demon.
W.W.: How do these lie demons work in the world after they have been created?
Etschewit: They move around and destroy what they can.
They destroy good thoughts and work against the development of the world. They try to destabilise the balance in favour of the adversary powers.
W.W.: Can these lying demons also influence people in this way,

Sphere 14 - Exusiai – 'Powers': protecting the heavenly spheres from all negative influences of the earthly dimension. Keep the world in balance and especially balance with the dark forces. Directing the regional Angels; they are spirits or creators of the forms. They are angelic beings in charge of the creation phase of every plant and tree. They are entrusted with creating and passing on the energy blue prints operating in the light ether. These blue prints are the basic structures of energy (lines, patterns, shapes) along which the elementals (gnomes, undines) build the physical. They received the impulses from the sphere 21, the trinity level and thus from the polarity of the creator and the earth-goddess matrix. Elohim

'The Transmitters'
Sphere 13 – Kyriotetes - 'Lordships': regulate the duties of the classes of angels beneath them; their energy is pure grace; mediators of the teachings of Christ, vitality and joy of life, spirits of wisdom; also in this sphere are for example Ignacio de Loyola, St. Francis of Assisi, St. Teresa of Avila, and St. Hildegard.

Sphere 12 – Archaï - 'Principalities': they guide the earthly regents, leaders, peoples, communities and represent thinking, spiritual energy and consciousness. They also act as the guardians of the 12 divine squares in cult places. I observed the temporary presence of an Archaï in a cemetery during a burial.

Sphere 11 – Church angels and archangels
Church angels (protection of altars, statues, crosses, objects), as well as the archangels: Raphael (healing), Uriel (teaching), Michael (bringer of light), Ezekiel (messenger of the divine will) Gabriel (archangel of creativity and bringer of good news) Jehudiel (service to others), Sandalphon (sound and prayers), Raguel (justice), Raziel (wisdom), Binael (personal trans-formation), Ramiel (hope & aspiration), Saraquiel (faith and spiritual power)

that these people speak new lies as a result?

Etschewit: Yes, of course. They can even become so big and strong that they possess people. There are so-called demon exorcisms in history and in the descriptions of the Bible, and they are very real....

Spectres

W.W.: When a community of people spreads many lies over a period of time, do the created lying demons then remain within the perimeter of this human community?

Etschewit: When a community of people speaks lies, indulges in certain lies, greater demons are created. These beings are called spectres. If these spectres are not dealt with by the human community through positive thinking, they come back to the community again and again, and they will meet people after death and even at their next birth.

W.W.: Suppose one creates a lie demon or a spectre.

How can one get rid of them again?

Etschewit: If one has lied by mistake, then it helps if that one

'Earth Angels & large Elementals'
Sphere 10 – Planet Earth Angel
The planet earth angel has its anchoring point in a remote area in Eastern Siberia, where there are practically no towns nor roads.

Sphere 9 – Large elementals of *earth, water, fire and air.*
The very small elementals are described under sphere 3. The large elementals are on a higher level in terms of knowledge, experience and competence, sometimes covering geographical areas as big as one fifth of France. This is similar to comparing a top manager and the plumbers and electricians working in the factory. Part of this sphere are also Angels of large range of mountains (Ural, Alps, Andes, Himalaya)

Sphere 8 – Angels of Science and Education Of oceans and smaller mountain ranges; Saint Bridget, the ancient goddess of healing, Spring, poetry and mountains (Brid in Celtic is Berg/Brig/mountain)

Sphere 7 – Angels of continents, Dakinis
The Dakinis are the Himalayan cousins of the Celtic Dagdas. They are comparable in their immense knowledge and wisdom. Beings from the Elf family are also part of this sphere. They are also participating in activating the Hartmann-Antenna.

Sphere 6 – Angels of Nations
Also works of art that are inspired by the Angel of the Nation and contribute to shape and evolve the spirit of a nation.

Sphere 5 – Sophia, Elementals of the 5th kind, Angels of large rainforests (Amazons, Congo, equatorial Africa), **big Elves.**
C suggests mentioning the following beings that also belong to Sphere 5 - the group souls of migrating birds, e.g. of wild geese, which some native Americans call 'The dream of the Earth'. Migrating birds carry thought structures or whole thought currents from one continent to another and as such are very important for

recanted immediately afterwards. If one
adds a good thought after such a lie, then
the relationship balances itself out
and the demon dissolves again.
With spectres it is much more difficult,
because it is very difficult to have a
community of humans agree
on the same positive
aspects. That is why spectres are much
more difficult to dissolve.
The human I's are free, and if today a
community of humans
create a spectre together through a lie, it
is difficult for the free
I's to come together again the next day
under the opposite sign.
But if some people from this community
endeavour, then positive beings can also
emerge, who then enter
into a fight with the
spectres. The spectres and
the struggles with them
burden the earth karma, the whole earth
environment, and at present there are
many such battles, which also create
thunderstorms. In these battles
the Michaelic hosts are working on the
good side.

the brotherhood of mankind. Also belonging to this sphere is the exchange of new cultural ideas of thoughts through books and cultural exchange. **Sophia,** who can be seen as the sum of cultural heritage and wisdom of the people and animals of the earth also belongs to Sphere 5.

Elementals of the 5th kind

According to C these elementals began to function in recent years (around 1993). Their task is to encourage and further the cooperation between humans and nature spirits.

They are found especially where cooperation between humans and nature spirits is taking place, and there are relatively few in existence. Colleagues in Dordogne have been cultivating herbal teas and spices for over 25 years with a special attention to include all concerned nature spirits. Towards the end of their activity, they had six elementals of the 5th kind on their plantation. Due to my ongoing communication work with nature spirits, there is also one behind our house. They explain to nature spirits how humans function and what their intentions are. This helps angels, in for example smaller rainforest areas (Indonesia etc.) to build up trust in some humans in their vicinity. Many nature spirits are wary of humans and do not understand why humans do not respect nature or communicate with them.

'The Servants of Nature'

Sphere 4 – Shamanic Light beings are astral beings with great wisdom operating between different realms (nature spirits, animals, insects, beings of the angelic realm, intrusive forces). As light beings, they are working towards keeping the spiritual dimensions operating within their field of action.

Angels of smaller rainforest areas   (Indonesia etc.)

Sphere 3 – Nature Spirits, Regional Angels and Kali

Regional angels have a 50 km radius of action. They are the superiors of Landscape angels.

Memory beings 21 April 2022

There are memory beings (MB). They are part of the upper layer of the life ether. There are different types: MB of angel activity, MB of human thought activity, MB of elementals activity,

C has guided me to sense energy traces of buildings that no longer exist, such as the former cathedral of Luxembourg or the Basilica in Paris on the Ile de la Cité [4], where archaeological evidence of it has been found under the 'Place du Marché aux Fleurs'. I can sense the walls, the altar and the magic squares. Knowing that every manifestation is connected to a spirit, I ask C what kind of spirit this is.

Are there special memory beings? - yes
Do they live in the upper life ether? - yes
Are there different types of memory beings? - yes
Memory beings of angelic activity, as in magic squares? - Yes
Memory beings for the activity of elemental beings? - yes
Are there memory beings for the traces of human activity? - yes
They are neither 'spirit beings' nor 'astral beings', but what I call 'beings with limited sentience and thought'? - yes
Have I understood you correctly? - yes
Thank you very much - it was a pleasure
A = Square of the Church Angel
BV = square of the Black Virgin
HS = magic square left of the entrance
C = magic square left altar corner
Autel = altar

Nature Spirits, in the narrow sense of the term, are the **Elementals**. The very small elementals are the workers on a molecular level and as such are the smallest spirit beings. In this sphere No 3 we find the four elementals linked to one of the four elements of creation: earth (gnomes), water (undines), fire (salamanders), air (sylphs). The 5th kinds of elementals are part of sphere No 5. The large and very large elementals are described in sphere 9.

Kali is the Hindu goddess who is considered to be the master of death, time and change (dissolving and renewal). She is called Kali Mata 'the dark mother'. These descriptions make her identical to the Black Virgin or Virgin Blackness. On the sacred site nearby she is one of the 'four queens': Black Virgin (north/Winter/New Moon), Kali (east/autumn/waning moon), Brid (west/Spring/waxing moon), Sophia (south/Summer/full moon)

Angels of larger deserts (Gobi, Sahara, etc.)

Sphere 2 – Angels of small lakes, rivers and deserts

Sphere 1 – Landscape angels, Devas, Dagdas.

Landscape angels have a radius of action of approximately 1 km. They advise nature spirits and Devas about all 'daily' questions concerning the functioning of nature. They themselves have access to elders (Regional angels) for any questions concerning the balance and harmony in their territory. These daily matters can concern the cohabitation of different species (animals, insects, invasive varieties of plants) and unusual changes in climate, flooding, drought, intrusive beings, lack of pollinators, etc.

Devas are beings in charge of the communication between the small elementals and the angelic realms (here especially the landscape angels). Devas are neither elementals nor part of the angelic hierarchies. They form a hierarchy of their own with princesses and queen Devas. Devas are part of the astral world. There are Devas with different types of tasks.

The background to the Corona virus situation is still relevant today, in April 2022, and may remain so for a long time. They points out weaknesses in our society.

Auragramm of C of 21.3.20 on the Corona Virus situation. I ask C, Collegium of Spirit Beings of Rocamadour, for their point of view and am using for this the form of the Auragramm. We put humanity in the centre of the Auragramm. Around this we find the energy envelopes of the aura [6]. I ask to be shown the energetically important places of the situation. These are mostly spirit beings, adversaries or egregors. The sequence order is important. Each point shows a clear accumulation of energy, which I locate with my hand and the Hartmann antenna.

1. At the point of the divine soul they showed me a collective thought-form, an egregor. This accumulation of thought structures and emotions has been building up for over 400 years and concerns our materialistic view and relationship to the earth and nature. This shows a lack of respect and has directly caused the virus.
2. They then showed me the root chakra, which is our relationship to the earth, to nature and to our body and our physical circumstances, which need healing.
3. I was then shown that at the heart chakra, a fundamental transformation towards more empathy urgently needs to be undertaken individually and collectively.
4. The last thing they pointed out to me was an adversary being in the etheric blueprint layer. These spirit beings from Redshift 7, a galaxy, point us to the ultimate illusion of being able to shield ourselves from other people, countries, nature spirit beings, viruses, etc., by means of walls, masks, prejudices, artificial boundaries and the like, as long as causes 1-3 are not resolved. The fact that these beings are in the energy layer, which contains the blueprint for the etheric field requires an explanation. Our erroneous way of thinking (egregor under 1) has caused a weakening of our immune systems. It is the result of points 1-3 and can only be healed by solving these three problem areas - individually as well as collectively - i.e. by changing our attitude towards the Earth and Nature.

Dagdas: I first came across a Dagda in connection with the quartz crystal I use. He regulates the communication aspect of the crystal, facilitating the communication lines of any spirit being that needs support from humans. As all spirit beings, Dagdas have great knowledge and wisdom concerning their specific field of action. Dagdas know a lot about the energy connection through crystals and how to establish the connection with other spirit beings. Their influence covers a large territory and they cooperate with Dagda colleges about the establishing contacts with specialists on their territory. This can be a human (with or without crystal), tall Elves, beings of the magical realm, etc. who may be able to help answer a specific question.

The tall Elves are about 3 m tall. They are the intermediaries who help regulate relationships between groups of beings when changes are happening. They are also carriers of beauty, nobility, courage, equality, hopes and aspirations. There are male and female Elves.
The **small fairies** have their gathering places in 'fairy bushes'. These fairies are less than 10 cm tall, and are bearers of joy and lightness, often creating an atmosphere of playfulness, in a sunbeam for instance.

Communicating with nature beings needs to be built up as a practice. Mediators have to be found and trained. The present change of attitude towards nature spirits will bring results when they feel that we begin to respect them. Greeting them is always a useful first step as this shows our change of attitude. The rest will follow surprisingly easily, even if we cannot quite imagine how. They will help us. That is my experience.

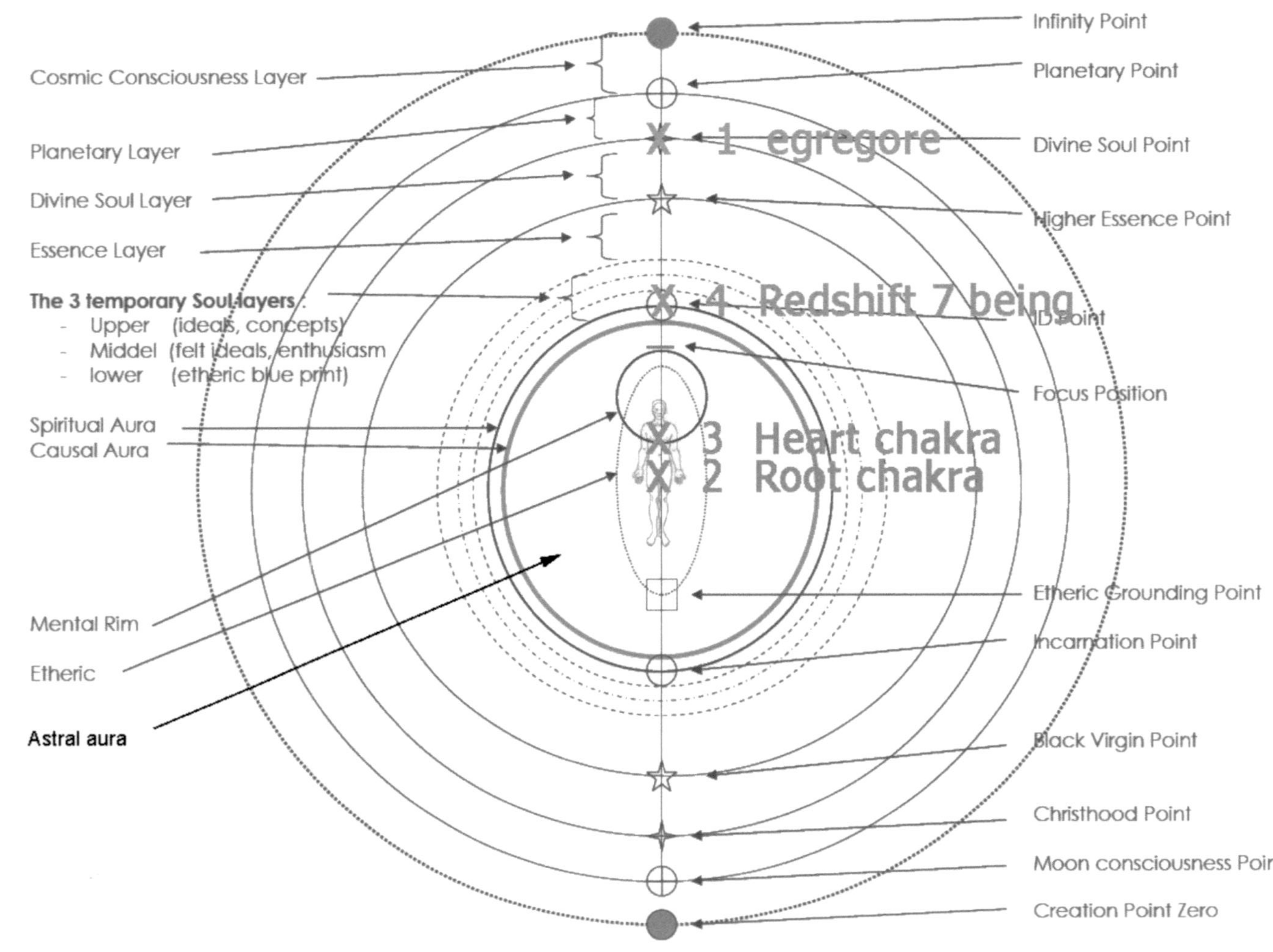

Daniel Perret – Beings of the Invisible Dimensions
Left side: Conversations with beings

Page 14 I present an interview with a large water elemental of the Loire River system in France as I wish to share an example of a communication that I have experienced. I had no mandate from a citizen group to do the interview, so there was no political aim. The interview concentrated on what the large water elemental wanted to share on that particular day. It taught me about the geographical extent of his river system, which included the estuary into the sea. The interview shows that he has constant access to all the details in his vast geographical area.

Being physical we can do things they cannot, for instance we can send a prayer using our thoughts and feelings. We seem to have a contact to the spiritual realm that is particular to human beings. Although scientific research has shown the positive effects of prayer, it is obvious to me that we still know very little of the power of our higher self, our thoughts and prayers. I came to understand that perhaps, because we are physical, we could establish connections between nature spirits and specialised beings of the divine field, that nature spirits for some reason cannot contact themselves anymore. There are more aspects that became evident in using distant healing for nature: Over the years nature spirits have explained how they function, the kinds of problems they are facing and how they are organised. [16]

There are different types of communication with invisible beings, but perhaps only one way: that of staying true to our own feelings and following our own path and copying someone else. The simplest and perhaps most valuable way of communicating is by following the feelings in our heart, as feelings are the essence of all communication. We should always remember that consciousness is primarily based on feelings and the heart, so that no thought, no words are really necessary. Many will only have a heart contact with invisible beings in a first phase.

Crystal communication lines A few years ago, the first energy line appeared close to the crystal that is on the floor in our meditation room. This surprised me, I hesitated for several days then I followed the line in order to understand why it was there. It led into our forest and to my first encounter with a nature spirit being. After that, hundreds of these lines and structures gradually appeared. Different segment angles mean different types of nature beings or landscape angels. They all point in the geographical direction in which the being has its anchoring place. A specially commissioned crystal being, a Dagda, takes over the organisational side and brings beings and lines to me. Angelic beings help him with this. This 'system' is very unlikely to be imitable, but it hows how inventive and precise contact with spirit beings can be. Some of my students also work with a crystal and a Dagda.

I describe elsewhere how all these lines are requests from beings that know that I can bring them what they need with distance healing. How this works I can only guess at for the most part and have described elsewhere. [5]

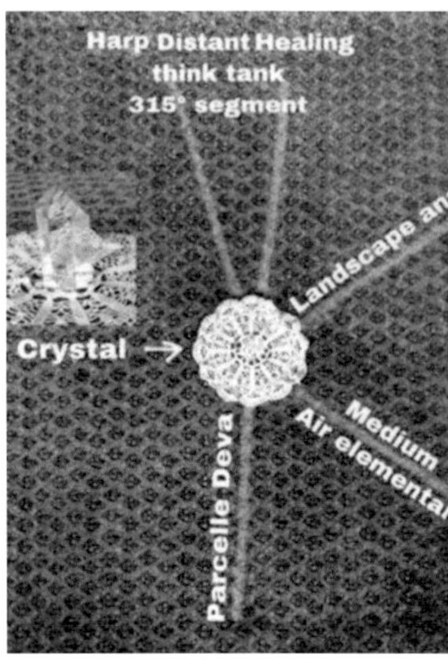

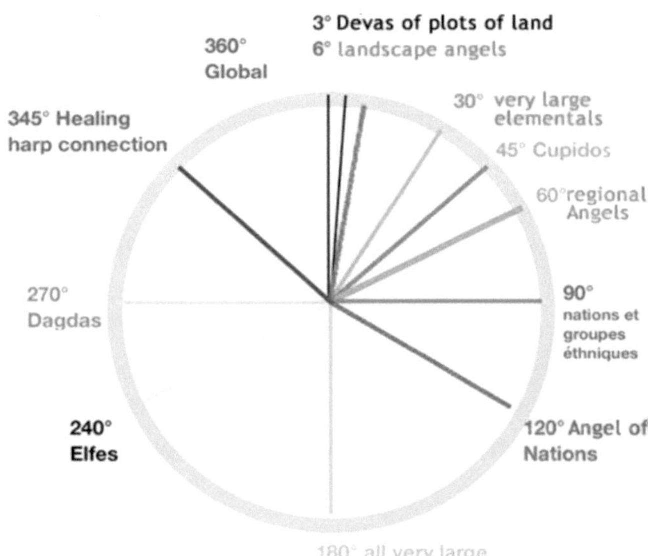

Sincere generosity does not expect anything in return. We can thank them, greet them, send distant healing, tell them when we find their tree, flower, landscape beautiful. A more explicit contact with them will happen, but it takes time, a building of trust.

Some people are able to talk to nature spirits or hear clear sentences. Other people, like me, have to work on finding the right questions to get clear responses such as yes or no. Once or twice I have received information in dreams when my teacher has communication with me. I feel energy columns in nature or in a room, which always correspond to a being. Then I have to ask with my Hartmann antenna who it is, if they are well or whether I can do something for them, and so on.

Our contact with beings of the divine field must remain practical and useful. There are very valuable issues where spiritual beings can help us to improve our daily life in a community. The following sections have all been prepared in collaboration with the spiritual Think Tank C. I use my Hartmann antenna to locate the points on a map and get the yes/no answers. Everyone can find their own way to feel energy (pendulum, hands, intuition, etc.).

Village and neighbourhood angels
We usually speak of a lively village spirit, when this community is particularly culturally and socially active. The village spirit is more than an abstract concept as we actually find an angel, mostly in the geographical centre of a village or district, who brings down divine Inspirational energy and keeps it ready. It depends on the actions and attitudes of the community members, how much of it really flows into the community and 'inspires' it. These angels then also feel more or less 'useful'.

The decisive factor is how the residents respect these beings and cooperate with them, even sometimes ask them for help. Since these beings respect our free will, they do not intervene

As the great elemental of the Rhone River,
I am what can be called the spirit of the
Rhone. My role is to advise and assist the
myriad of medium and small water elementals
down to the undines. I know everything that is
happening at any given moment in my entire
river system or watershed as you call it. My
anchor point is on the island of the Round
Table south of Lyon.

In your statement* it says:
"Establish the primacy of water as a universal
resource essential to our survival and that of
future generations."
All living things form a community of sentient
and conscious beings. Water and its
elementals have no particular priority, for life
depends on the actions of all. As a spirit of
the Rhone, I am only an actor among the
spirits of nature.
We are pleased with your initiative and
interest in the Rhone. To give a status and a
legal personality to this waterway is certainly a
good idea. But it is not enough.

A river system is only the material aspect. The
fundamental problem was created when the
humans of the watershed forgot that every
tree, every hill, every forest and every river
have conscious beings that take care of them.
By considering us as purely physical objects
they have lost respect for the sentient being
behind the physical aspect, as a delegated
being of the Heavenly Father and the Earth
Goddess,
the Dea Mater, Demeter.

We have been waiting for centuries for our
cooperation to resume - between you humans
and us nature spirits
*) L'Appel du Rhône [15]

until they are asked. In most of the villages that I have studied in our area, only an average of 30% of the energy that angelic beings bring down, is used by the community. That is not much. The percentage can vary from day to day and seems to depend, among other things, on the following:

- Respect of nature (e.g. level of pollution).
- Respect of animals (pets and wild animals)
- Degree of cooperation with devas, nature spirits and angels
- Prayers or appreciative words that are spoken or thought daily
- Faith lived by community members
- Respect and compassion shown to fellow residents
- Gratitude toward creation
- Respect and care towards our own bodies
- Caring for the spirit of the village (culture, festivals, etc.)
- The care of the beauty of the village or district

Angels of Nations
Some tasks of the angels of nations, acting as advisers for:

- regional angels and in general all the light beings that are
  are active on their territory
- the angels of ethnic 'minorities
- the Deva Queens of the Territory
- the great river systems and lakes with their own
  nature spirit beings
- the adjacent seas
- the air, mountain, water and earth elemental beings
- the dialogue with the great elemental beings concerning
  pollution and environmental stresses of all kinds
- the symbols of the nation: anthem, flag, monuments, culture...
- the language(s) of the nation

- the treatment of minorities, the poor, etc.
- relationship with other countries (treaties, cooperation, agreements, friendly visits, refugees, etc.)
- the state and development of governance
- political development (degree of development of democracy, dialogues between groups, etc.)
- the spiritual development of the country (religions, places of worship, development of spiritual values)
- empathy, respect, equality, justice system, peace
- the essential human values such as free will, creativity
- the country's contribution to world peace, progress, especially with regard to the continent to which the country belongs to
- the management of crises (unemployment, social unrest, depression, collective fears, spiritual blindness, discouragement lack of vision for the future, etc.

Hectograms of communities

The form of the heptagram (*hepta* meaning 7 in ancient Greek) is to be understood abstractly like a geometric symbol. These heptagrams can be used to evaluate the degree of cooperation between the visible and the invisible world at a given moment. They could be labelled as socio-spiritual, linking the spiritual with the quality of the social interaction of a community. These values can be compared at different moments in order to see how events or actions have affected a community. Angels are primarily bringer of divine inspiration. Their purpose is to improve this cooperation. As with all heptagrams, this opens the possibility of evolutionary work with a community, be it a neighbourhood, a village, a school, a business, an association, an agricultural cooperative or a country.

In all 7 places of this cooperation heptagram we find Kyriotetes Angels (Sphere 13) as well as No. 5 Elementals. The latter appeared worldwide only in 1993 and have the task to support the cooperation between nature spirits and humans. Both, Kyriotetes as well as these elemental beings, help to use the cooperation heptagrams. To determine the situation of an organism, each field 1-7 can be interrogated directly by means of pendulum, Hartmann antenna or the like. Each of the seven points can give information about the degree of potential utilisation of this organism at a certain point in time, e.g. expressed in percentages. Each point can be further refined and adapted to the needs of an organism or place.

The heptagram structures (the seven circles and connecting lines) can be found at each examined place or organism on a map or on the spot. To query the values, however, this is not necessary.

The 7 energy centres or main themes of a heptagram:
1. goal orientation, awareness of the spiritual task
2. acceptance of the present circumstances
3. knowledge of the beings involved, visible and invisible
4. wisdom, which is currently being applied
5. ability to relate, quality of communication within the community
6. lived joy of life in the community
7. implementation of the inspirational energy of the spirit/angel of the community

Heptagram of Paris

For Paris, within the 1st ring road, the numbers on 21/11/2018 (before the 2018 riots) > and on 15/4/2022 are:

1: 79%>79%

2: 45%>87%

3: 25%>48%

4: 51%>39%

5: 60%>48%

6: 82%>57%

7: 83%>87%

The comparison allows exciting differences to appear and invites closer examination and interpretation. The high values for 1 and 7 can be explained by the special position of Paris as a well-known tourist city, the care given to its monuments, theatres, buildings and parks, and the awareness of possessing a unique and intact cultural heritage.

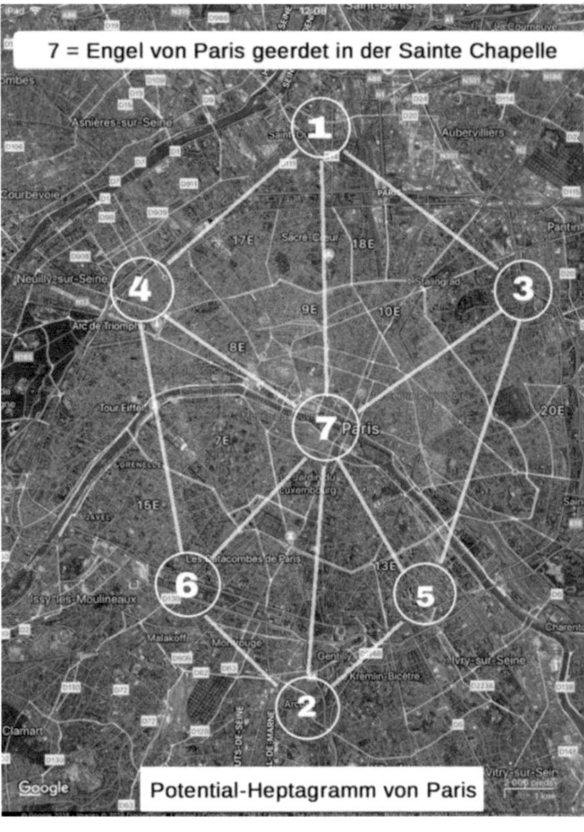

In the heptagrams the places 1-6 are always connected with the city angel 7. It is my observation, position 7 is always centrally and very meaningfully chosen in all heptagrams. The city angel of Paris is anchored in the 'Sainte Chapelle', not far from Notre Dame. All locations 1-6 are also connected to each other in a ring shape. This is not coincidental. There is a feeling connection between the centres thus connected:

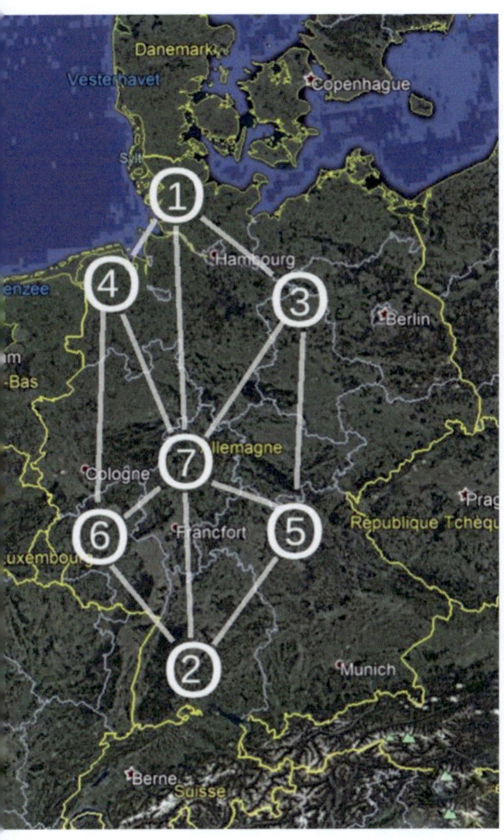

1 is future-oriented, 2 is the present starting point, 7 is the centre, the spirit being or angel of the landscape. 3 and 5 are tending towards the masculine, the expressive side, 4 and 6 are on the feminine side. 1, 3 and 4 belong to the spirit-mental part including the heart, 2, 5 and 6 to the lived part. 1-6 are in the case of Paris accompanied by sphere 11 beings: Archangels.

Continent & Country Heptagrams, At points 1-6, are sphere 15 angels 'Dynameis'. While **the angel of Germany** (at point 7 as always) is a sphere 6 angel. Because of the high abstraction level, all continent and country heptagrams are North-South oriented even for countries like Austria or Russia, which have rather an East-West spread. (All this information comes from C.) The decentralised position of the heptagram of Germany points to an incomplete integration of the eastern part of Germany. There are obviously further political, cultural, social and geomantic healing measures necessary.

While the **Europe Angel** is located almost in the geographical centre of Europe, the heptagram of Europe does not clearly reach the south of Europe, apparently because of the missing integration of the Balkan states. Interesting how point 6 'lived joy of life' is situated in Paris. There is much more information in the details.

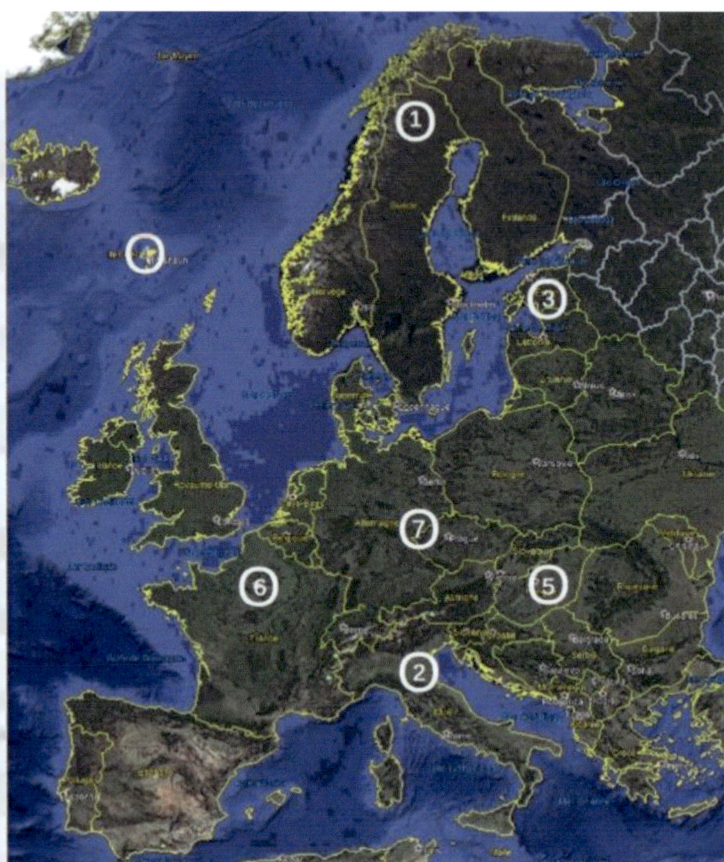

I would like to conclude by looking at the 'geomantic graffiti' aspect. C explains: the virtual heptagram did not exist before I asked the question about it. When C showed it to me on Google Maps, energy imprints appeared simultaneously in the landscape in question. I find the same seven rings with a meter diameter each as well as the energy columns of the spirit beings when I find them outside.

While these energy rings had originated from my thinking (thus 'geomantic graffiti'), the Kyriotetes and nature spirit beings to be found there had not been created or brought there by me. There would also be a symbolic, virtual aspect, in that each time such a cooperation heptagram would be created by people, they would also participate and be present accordingly. The heptagrams leave the connecting lines between the seven circles on the spot. The cooperation of the visible and the invisible world is a mystery and will probably remain so, even if we can begin to learn about this subject.

The Spiritual Organisation in a Landscape

Three overlapping networks of beings operate simultaneously in the same landscape. What we call energy lines or energy structures are always manifestations of consciousness:

1) **Landscape and regional angels dealing with their nature spirits**: Devas and elementals, including for instance big elementals of parts or whole river systems. These places are all linked through energy lines and are the inspirational dimension bringing impulses from the divine field. Landscape angels deal with a territory of 1km radius, regional angels with territories of 50km radius.

2) 2nd network is **linking the sacred places** of the landscape through energy lines. Examples Dordogne (SW France): Berboules (Sergeac), Côte de Jor Tibetan Buddhist sacred place (St. Leon s/Vézère), hill 500m SE from Roc St. Christophe at Peyzac Le Moustier, Puy de Pauliac 1200m NE of Aubazine (Corrèze), hill 250m West above Grotte Bernifal (Les Eyzies), hill 800m NE of Château du Mas Nègre (Valojoulx) – these places are looked after by high angels (Thrones), protecting and maintaining sacred energy grids and places.

3) 3rd network: the **Guardians of the Landscape or 'the Wise Ones'.** They are responsible for a geographically defined area, a felt unit in a landscape. These beings are non-incarnate, non-human beings, created by the Mother Earth Being. They are a different type of spirit beings than the nature spirits, Devas and elementals. They have the overview of all aspects of quality: water, pollution, soil, vegetation, animals, birds, air, mountains, hills, micro climates, etc. Their complete knowledge of the landscape and its history is based on clairvoyance and wisdom. They also are all linked with each other through energy lines.

The relevance of Auschwitz today

2018 I asked C about the degree of healing reached through all the prayers in and for Auschwitz since WW2. I was astonished by their response.

They answered: 'Barely 12%!' and went on to explain that a complete healing of the energies and memories concerning the Auschwitz-Birkenau concentration camp would only happen when the people of Europe truly understood what fascism was and fully rejected this ideology.

On Light and Dark   Continued from page 49

Light beings and Beings of the Dark need to stay in a balance.
On a spiritual level a opposite force always appears when
something is created. When a tree seedling comes to life a new tree
spirit is created at the same time as a decay being, which will take
care once that tree, or a branch, is in a dying, descending phase.
Problems only occur when dark beings get the upper hand. This
often happens because of human shortcomings - that is lack of
purity and light in actions and thoughts.

Creation happens when Light penetrates Darkness, when the Father
principle – loving intelligence – meets the Virgin Blackness, the
Black Virgin – intelligent love. In that process the darkness is not
pushed away, only complemented by its other polarity.

The Father impulse or Universal Intelligence creates both types of
impulses: for light beings and for dark beings. The Mother impulse
accepts both and creates light and dark beings. Good and bad do
not exist as such, they are complementary polarities, and are both
necessary for evolution. Balancing these opposites comes about
through strengthening the light and through respecting both
polarities as ultimately of divine origin.

Light beings are on an ascending path to the Light.
Dark beings are on a descending path towards decay and the dark.

Julia Wright, from the University of Coventry [16] on biodynamics and
indigenous traditions:

1. Humans and nature are one.
2. Everything is alive, no dead and alive elements.
3. There is a constant search for balance.
4. Cycles of change happen naturally and death brings new life,
   rather than claiming the dying earth must be saved by us.

The reflector or warmth ether is one of the four states and layers of the etheric energy field around our bodies. Every physical manifestation, like plants, trees and the earth as a whole, has these four layers of the etheric. The warmth ether is the outermost layer, that is, the layer furthest away from the physical body. This warmth or reflector ether has three sublayers. The outermost layer is the interface giving us access to nature spirits, to universal wisdom and inner peace.

The extent and depth of my collaboration with spiritual beings of all kinds is almost impossible to describe. For one, I am not always aware of when they have already played a decisive role in the formulation of a question. They have also repeatedly pointed out errors of thought in the text to me and helped to correct them. I had written that 'a being that has senses is also (meaning generally:) sentient'. C corrected me to write the present version. C tells me that most of my communications are with them, unless I am speaking directly to a particular nature being. With the help of C, we have published ten books in the last few years. I feel it is important to pay due tribute to them here and present them in detail elsewhere. [2]

Finally, a communication from a great elemental being of the 5th kind, who specialises in co-operation between nature spirits and human beings, and has asked me to add this:

"It is possible for every human being to communicate with nature spirits. Humans can start simply by asking them a question about some aspect of nature. Asking a question is the most important step because it shows that one respects the other, shows interest and is open to consider their existence, their intelligence and their opinion. Each person can get an answer with their own means: with their feeling, intuition, a vision, a radiesthesia tool or a more sophisticated channelling consisting of whole sentences."

A heart felt communication always connects to the whole, to beings of the divine field, and to our inner feelings.
The three are inseparable.

Daniel Perret – Beings of the Invisible Dimensions
Right side: continuous explanatory text

References

My books are best ordered at Books on Demand (BoD) bookshop, see www.vallonperret.com or on Amazon

1) Earth-Healing   (in German and Frensch only)
2) The Intelligence Behind Nature
3) Music as a mystical journey
4) The Lost Orientation of Cathedrals   (in French and German only)
5) The Harp in Distant Healing
6) Science of Spiritual Healing II, A Wider Self
7) L'Accès aux Mondes invisibles   (only in French)
8) The 12 Magic Squares as Divine Seals - Perception and Interpretation of Subtle Energy   (in German only)

Books by other authors
9) J.M. Keynes, https://letstalkaboutbooks.blog/2020/10/21/newton-and-alchemy-i-john-maynard-keynes-and-the-myth-of-newton-the-magician/
10) Michael Newton, The Journey of Souls, Llewelllyn Publications, 2002
11) Christ Letters, free download at: www.christsway.co.za
https://thechristletters.weebly.com/uploads/1/2/5/7/125746987/christ_letters_original_version.pdf - printed book at
https://www.christsway.co.za/
12) Federico Faggin, Wall Street International Magazine, series from 11 December 2020 - 11 April 2021: The Nature of Consciousness.
13) Flensburger Hefte No. 79 'Nature spirits and what they say',
Floris Books, 2004
14) www.oca.org/saints/lives/2022/10/03/102843-hieromartyr-dionysius-the-areopagite-bishop-of-athens
15) L'Appel du Rhône
https://www.appeldurhone.org/_files/ugd/6963cc_274fe3817316413e9135f89144352232.pdf
16) J. Wright, N. Parrott, subtle Agroecologies – farming with the Hidden Half of Nature, Taylorfrancis.com, 2021
Eva Høffding's website: https://ignatiushr.dk/
Webpage Wolfgang Weirauch: www.wolfgang-weirauch.de

We are a shiny thread in a fantastic tapestry
of colour, sentient beings and beauty.

Daniel Perret - www.vallonperret.com
danielperret.bandcamp.com